PERSPECTIVES ON COMPLEX GLOBAL CHALLENGES

PERSPECTIVES ON COMPLEX GLOBAL CHALLENGES

Education, Energy, Healthcare, Security and Resilience

Edited by

ELISABETH PATÉ-CORNELL
Stanford University,
California, US

WILLIAM B. ROUSE
Stevens Institute of Technology,
Hoboken, NJ, US

CHARLES M. VEST

For general information on our other products and services or for technical support, please contact our
Customer Care Department within the United States at (800) 762-2974, outside the United States at
(317) 572-3993 or fax (317) 572-4002.

Wiley also publishes its books in a variety of electronic formats. Some content that appears in print may
not be available in electronic formats. For more information about Wiley products, visit our web site at
www.wiley.com.

Library of Congress Cataloging-in-Publication Data applied for:

ISBN: 9781118984093

Set in 11/13pt, TimesLTStd by SPi Global, Chennai, India.

Printed in the United States of America

10 9 8 7 6 5 4 3 2 1

TO CHARLES M. VEST

FRIEND, MENTOR, AND INSPIRATION

To his colleagues, his friends, and especially the faculty and students of MIT, of which he was the President from 1990 to 2004, Charles Vest was the ultimate role model. His intellect, his kindness, and his fairness have changed the professional and the personal lives of many around him. His death, in 2013, has been an immense loss, first to his family, and also to the world of academia, to the National Academy of Engineering that he led from 2007 to 2013, and to both of us, Elisabeth Paté-Cornell and Bill Rouse, who had been working with him on this book. The three of us had the vision of gathering the thoughts of a few of the luminaries among our friends who could help us think through some of the most difficult problems that we are facing, and that our children will most likely face as well.

We wanted to pursue this effort to completion, acknowledging Chuck's collaboration with us and hoping that by doing so, we can honor his memory. So we dedicate this book to Charles Vest, and also to his family, and especially to his wife Becky with our respect and our affection.

CONTENTS

CONTRIBUTORS

Norman R. Augustine served as CEO of Lockheed Martin Corporation from 1995 to 1997, following having been CEO of the Martin Marietta Corporation from 1987 to 1995. He served as Under Secretary of the Army from 1975 to 1977. He has been Chairman of the American Red Cross and Chairman of the Defense Science Board and is a former member of the faculty of Princeton University. He is a member of the National Academy of Sciences and former Chairman of the National Academy of Engineering.

Lawrence S. Bacow served as Chancellor of MIT from 1998 to 2001, as the President of Tufts from 2001 to 2011, as a trustee of Wheaton College from 1998 to 2008, and since 2011 has been a member of the Harvard Corporation, the senior governing body of Harvard University.

Craig R. Barrett served as CEO of Intel from 1998 to 2005, as well as Chairman until 2009. He is a member and former Chairman of the National Academy of Engineering. He currently serves as President and Chairman of BASIS School Inc., a charter school group.

Michael Batty is Bartlett Professor of Planning at University College London where he is Chair of the Centre for Advanced Spatial Analysis (CASA). Before his current position, he was Professor of City Planning and Dean at the University of Wales at Cardiff and then Director of the National Center for Geographic Information and Analysis at the State University of

New York at Buffalo. He is a Fellow of the British Academy (FBA) and the Royal Society (FRS), was awarded the CBE in the Queen's Birthday Honours in 2004, and the 2013 recipient of the Lauréat Prix International de Géographie Vautrin Lud.

Denis A. Cortese is Foundation Professor, Department of Biomedical Informatics and Director, Healthcare Delivery and Policy Program at Arizona State University. He is also Emeritus President and Chief Executive Officer, Mayo Clinic and a member and former chair of the National Academy of Medicine Roundtable on Value and Science-Driven Healthcare, as well as the Healthcare Leadership Council.

John Deutch is Emeritus Institute Professor at the Massachusetts Institute of Technology where he has served as Chairman of the Department of Chemistry, Dean of Science, and Provost. He is chair of the Secretary of Energy Advisory Board, and a member of the National Petroleum Council and the Hamilton Project. He served as Director of Central Intelligence from May 1995 to December 1996. From 1994 to 1995, he served as Deputy Secretary of Defense and served as Undersecretary of Defense for Acquisition and Technology from 1993 to 1994. He has also served as Director of Energy Research (1977–1979), Acting Assistant Secretary for Energy Technology (1979), and Undersecretary (1979–80) in the United States Department of Energy.

Jacques S. Gansler is a Professor Emeritus at the University of Maryland School of Public Policy and he also is the Glenn L. Martin Institute Fellow of Engineering at the University of Maryland, School of Engineering. In addition, he is the founder and CEO of The ARGIS Group (Analytical Research for Government and Industry Solutions) and independent research and consulting firm. He served as Under Secretary of Defense for Acquisition, Technology and Logistics from 1997 to 2001. He was Executive Vice President and Corporate Director for TASC from 1977 to 1997. He served as Deputy Assistant Secretary of Defense (Materiel Acquisition) from 1972 to 1977. He is a member of the National Academy of Engineering.

Linda Darling-Hammond is Charles E. Ducommun Professor of Education and founding Director of the Stanford Center for Opportunity Policy in Education (SCOPE) at Stanford University. She was the founding Director of the National Commission for Teaching and America's Future, and served as chair of President Obama's education policy transition team in 2008.

Siegfried S. Hecker is Professor (Research) in the department of Management Science and Engineering at Stanford University, and a Senior Fellow of the Stanford University Center for International Security and Cooperation (CISAC) of which he was a co-director from 2007 to 2012. He was Director of Los Alamos National Laboratory from 1986 to 1997, a Los Alamos senior fellow until 2000 and co-recipient of the 2009 Enrico Fermi Award. He visited several times the Yongbyon nuclear facility in North Korea and reported its state of advancement to the US Congress in 2010. He is a member and a past councilor of the National Academy of Engineering.

Michael M.E. Johns is currently Professor of Medicine and Public Health at Emory University, where he served as Chancellor from 2007 until 2012, before which he served as Executive Vice President for Health Affairs, CEO of The Robert W. Woodruff Health Sciences Center, and Chairman of the Board of Emory Healthcare. From 1990 to 1996, Dr Johns was Dean of Johns Hopkins School of Medicine and Vice President of the Medical Faculty. He is a Member of the National Academy of Medicine (NAM) and served as a member of the NAM Council.

Henry A. Kissinger is a German-born American statesman and political scientist. He served as National Security Advisor from 1969 to 1975 and as Secretary of State from 1973 to 1977 in the administrations of Presidents Richard Nixon and Gerald Ford, pioneering the policy of détente with the Soviet Union, orchestrating the opening of relations with the People's Republic of China, and negotiating the Paris Peace Accords, ending American involvement in the Vietnam War. He is a recipient of the Nobel Peace Prize.

Michael E. Leiter was the Director of the US National Counterterrorism Center (NCTC), having served in the Bush Administration and been retained in the Obama Administration. Before joining NCTC, Leiter served as the Deputy Chief of Staff for the Office of the Director of National Intelligence and as the Deputy General Counsel and Assistant Director of the President's Commission on the Intelligence Capabilities of the United States Regarding Weapons of Mass Destruction. He was with the Department of Justice as an Assistant US Attorney from 2002 to 2005.

Herbert Lin is Chief Scientist at the Computer Science and Telecommunications Board, National Research Council of the National Academies. Before his NRC service, he was a professional staff member and staff scientist for the House Armed Services Committee from 1986 to 1990, where his portfolio included defense policy and arms control issues.

Elizabeth A. McGlynn is Director of Kaiser Permanente's Center for Effectiveness and Safety Research (CESR), a virtual center designed to improve the health and well-being of Kaiser's more than ten million members and the public by conducting comparative effectiveness and safety research and implementing findings in policy and practice. Dr McGlynn is an internationally known expert on methods for evaluating the appropriateness, quality, and efficiency of healthcare delivery and has led major initiatives to evaluate health reform options under consideration at the federal and state levels. She chairs the Scientific Advisory Group for the Institute for Healthcare Improvement. McGlynn is a member of the National Academy of Medicine.

Richard A. Meserve is the President Emeritus of the Carnegie Institution for Science and Senior Of Counsel with Covington & Burling, LLP. He previously served as Chairman of the US Nuclear Regulatory Commission. He is the Chairman of IAEA's International Nuclear Safety Group and Co-Chairman of DOE's Nuclear Energy Advisory Committee. He is a member and councilor of the National Academy of Engineering.

Lloyd B. Minor is the Carl and Elizabeth Naumann Dean of Stanford University School of Medicine and a Professor of Otolaryngology-Head and Neck Surgery. He is also a Professor of Bioengineering and of Neurobiology, by courtesy, at Stanford University. He was previously the Provost and Senior Vice President for Academic Affairs at The Johns Hopkins University and, before that, Andelot Professor and Director (chair) of the Department of Otolaryngology-Head and Neck Surgery at the Johns Hopkins University School of Medicine and Otolaryngologist-in-Chief of The Johns Hopkins Hospital. He is a member of the National Academy of Medicine.

Michal C. Moore is Professor of Energy Economics and Senior Fellow at the School for Public Policy at the University of Calgary and is a visiting Professor of Economics and Systems Engineering at Cornell University. He is a former regulator of the energy industry in California.

Elisabeth Paté-Cornell is the Burt and Deedee McMurtry Professor of Engineering at Stanford University in the department of Management Science and Engineering, which she chaired from its creation in 2000–2011. She is a past President of the Society for Risk Analysis. She served as a member of the President's (Foreign) Intelligence Advisory Board from 2001 to 2008. She is a member of the National Academy of Engineering and of several boards including InQtel, Draper Laboratory, and Aerospace Corporation.

William J. Perry is a senior fellow at the Hoover Institution and the Michael and Barbara Berberian Professor at Stanford University, with a joint appointment in the School of Engineering and the Institute for International Studies. He served as the Secretary of Defense from 1994 to 1997, Deputy Secretary of Defense from 1993 to 1994, and Under Secretary of Defense for Research and Engineering from 1977 to 1981. He is a member of the National Academy of Engineering.

Richard Reed is Senior Vice President, Disaster Cycle Services, at the American Red Cross. In addition to his current role, Richard most recently served on special assignment to the White House as Deputy Ebola Response Coordinator to support the coordination, management and leadership of the US Government response to the Ebola epidemic. Before assuming his role at the Red Cross, Reed served in the Obama and Bush administrations as Deputy Assistant to the President for Homeland Security, Special Assistant to the President and Senior Director for Resilience Policy, and Special Assistant to the President for Homeland Security and Director of Continuity. He has held positions in the Department of Veterans Affairs, the Federal Emergency Management Agency, and the General Services Administration.

William B. Rouse is the Alexander Crombie Humphreys Chair of Economics in Engineering at Stevens Institute of Technology, and Director of the university-wide Center for Complex Systems and Enterprises. He is also Professor Emeritus of Industrial and Systems Engineering at the Georgia Institute of Technology. He is a member of the National Academy of Engineering.

Richard Schmalensee is the John C. Head III Dean, Emeritus, and the Howard W. Johnson Professor of Management and Economics, Emeritus, at the MIT Sloan School of Management. He served on the President's Council of Economic Advisors and currently serves as Chairman of the Board of Resources for the Future.

George P. Shultz is a distinguished fellow at the Hoover Institution at Stanford University. He served as Secretary of State from 1982 to 1989, Secretary of the Treasury from 1972 to 1974, Director of the Office of Management and Budget from 1970 to 1972 and Secretary of Labor from 1969 to 1970. Before entering politics, he was Professor of Economics at MIT and the University of Chicago, serving as Dean of the University of Chicago Graduate School of Business from 1962 to 1969.

Brent Scowcroft is a retired US Air Force Lieutenant General. He was the US National Security Advisor under US Presidents Gerald Ford and George H. W. Bush. He also served as Military Assistant to President Richard Nixon and as Deputy Assistant to the President for National

Security Affairs in the Nixon and Ford administrations. He served as Chairman of the President's Foreign Intelligence Advisory Board under President George W. Bush from 2001 to 2005.

Robert K. Smoldt is Associate Director, Healthcare Delivery and Policy Program, Arizona State University and Emeritus Vice President and Chief Administrative Officer, Mayo Clinic. He has been active in the Medical Group Management Association and serves on the Board of Trustees of Catholic Health Initiatives.

William W. Stead is Chief Strategy Officer at Vanderbilt University Medical Center and McKesson Foundation Professor of Biomedical Informatics/Medicine at Vanderbilt University. He serves on the Council of the National Academy of Medicine, the Division Committee on Engineering and Physical Sciences of the National Academies and the Health and Human Services National Committee for Vital and Health Statistics.

Deborah J. Stipek is the Judy Koch Professor of Education at Stanford University. She was Dean of the Graduate School of Education from 2001 to 2011. She was a Professor at the Graduate School of Education, University of California, Los Angeles, from 1977 to 2000, a Congressional Science Fellow from 1983 to 1984 and is a member of the National Academy of Education.

Theo Toonen is Professor in Institutional Governance and Dean of the Faculty of Behavioral, Management, and Social Science (BMS) at Twente University since April 2015. He is former Dean of the Faculty of Technology, Policy, Management (TPM) at Delft University of Technology (2008–2015) and former Dean of the Faculty of Social and Behavioral Science at Leiden University (2003–2008). He has held various advisory positions to the government and also has been a member of the independent Dutch Advisory Board on Water (AcW) under the Chairmanship of his Royal Highness Crown Prince William of Orange (2003–2012).

Richard H. Truly is a retired Navy vice admiral and former NASA astronaut, flying the Space Shuttles *Enterprise*, *Columbia*, and *Challenger*. He served as the eighth NASA Administrator, the first commander of Naval Space Command, and as Director of the Georgia Tech Research Institute and the National Renewable Energy Laboratory. He is a member of the National Academy of Engineering.

Charles M. Vest was President Emeritus of the Massachusetts Institute of Technology (MIT) and President Emeritus of the National Academy of

Engineering. A mechanical engineer, he previously was Dean of Engineering and Provost at the University of Michigan, and was a trustee of several universities and nonprofit organizations devoted to education, research, and national security.

Alexandros Washburn is Professor of Design and Director of the CRUXlab at the Stevens Institute of Technology. He is a practicing architect who has served in all levels of government, most recently as Chief Urban Designer under Mayor Michael Bloomberg of New York City. He is the author of *The Nature of Urban Design: A New York Perspective on Resilience.*

INTRODUCTION AND OVERVIEW

ELISABETH PATÉ-CORNELL, WILLIAM B. ROUSE,
AND CHARLES M. VEST

INTRODUCTION

North America, Europe, and, perhaps to a lesser extent, other parts of the world are awash in debate, conflict, and often vitriol regarding how best to engender a healthy, educated, and productive population that is competitive in the global marketplace. The public–private enterprises engaged in providing education, energy, healthcare, security, and resilience are under scrutiny and stress.

These large-scale public–private systems are very complex and involve rich interactions of behavioral and social phenomena with designed and emergent physical and organizational infrastructures that are increasingly enabled by computer, communications, and other technologies. The range of stakeholders and their sometimes conflicting interests includes the public, industry, government, and in some cases international bodies. These stakeholders are not monolithic as, for example, various segments of the public or of industry may have divergent interests.

We have chosen to focus on five areas – education, energy, healthcare, national security, and urban resilience. These areas are laced with economic, environmental, political, and technology issues. They tend to be confounded by upside benefits and downside risks. Nuclear power is a good example.

Perspectives on Complex Global Challenges: Education, Energy, Healthcare, Security and Resilience,
First Edition. Edited by Elisabeth Paté-Cornell, William B. Rouse, and Charles M. Vest.
© 2016 John Wiley & Sons, Inc. Published 2016 by John Wiley & Sons, Inc.

Much subtler are issues that include immigration and job creation, or the near-term costs versus long-term benefits of prevention and wellness in healthcare. There are overarching debates concerning who pays, who benefits, and who is responsible for risks.

A primary hindrance in these debates is the lack of a shared understanding of these enterprises and contexts, including how well they function, the nature of the real challenges they must address, possible ways to proceed, and risks associated with these endeavors. The needed understanding should also include the geopolitical contexts associated with these issues. Unfortunately, we are bombarded with sound bites purported to provide this understanding, but these polemics often serve no constructive purpose.

The objective of *Perspectives on Complex Global Challenges* is to bring nonpartisan, well-informed wisdom to bear on these challenges. An impressive range of thought leaders, from various perspectives, both conservative and liberal, was recruited to carefully and thoughtfully relate what we know, what we do not know, opportunities for progress, possible paths forward, and risks associated with each challenge. From their reflections and suggestions, we synthesize the key trade-offs, risks, and opportunities in these turbulent times.

TRADE-OFFS, RISKS, AND OPPORTUNITIES IN TURBULENT TIMES

Many of the contributions in this book point to trade-offs, risks, and opportunities inherent to the situation of the United States, as well as Europe, in these turbulent times. Emerging powers and new types of conflicts challenge the country's position as world leader, especially in Eastern Europe, South East Asia and the Middle East. A global economy, with the inherent vulnerabilities of intertwined financial systems, threatens the stability of both developed and developing countries, yet brings prosperity to many. The US competitiveness, adaptability, and willingness to seize opportunities and to confront long-term problems in a proactive way are keys to maintaining that position.

Turbulences and Risks

The world has always been a turbulent place at different scales: cities, provinces, or empires. But in that respect, several characteristics distinguish this century from previous times. The pace of change has accelerated drastically in the last century. Globalization has tightly linked both risks and opportunities worldwide, and the instant reach of information has shaped immediate perceptions of what would have been ignored earlier, or seen as local problems. The instability of the situation implies a constant change that cannot be captured by a snapshot of the moment. But a few fundamental principles remain.

Economic Turbulences and Political Polarization After decades of expansion following WWII, a global economic downturn has hit most countries. Among other factors, a number of "bubbles" in global markets, tight links among banks worldwide, and their use of computer programs based on similar principles have caused wide swings in the world financial system. At this point, a slow economic recovery seems underway, but many countries are facing the need to decrease their public spending. Drastic choices thus have to be made exposing, in the United States, a political polarization seldom encountered in its history. On one side, there is a desire to maintain and enhance social programs and public spending. On the other, the emphasis is on the decrease of the government size, role, and budget, and an increase in the freedom of private enterprises with less regulation and reduction of taxes. While the two main parties have collaborated to various degrees in US history, there seems to be few key actors ready to bridge the political gap. As illustrated by this book's contributors, this clash of ideologies affects many aspects of US life, including healthcare, education, energy, national security, and the management of growing (or shrinking) cities.

A healthy ideological debate can stimulate productive discussions at a time when the United States seems to be on the path of a slow recovery. Yet, a certain paralysis has hit the country in its search for a balance between the prosperity of some industries and top income groups, and the living standards of the middle and the lower classes. The income gap has grown significantly, causing a deep sense of insecurity in the less privileged social strata and leading for a while to an "occupy" movement, which, while lacking clear objectives, revealed a profound discontent. The unemployment rate has recently decreased but there is still a certain reluctance to invest in traditionally labor-intensive manufacturing sectors. In the blue-collar labor force, unemployment has been rooted in part in the decline of the US education system, especially at the K-12 and vocational/technical levels, in spite of high costs and a number of reforms. At the same time, the costs of healthcare have skyrocketed for a host of reasons including incentive systems, the costs of new drugs and technologies, and an aging population. Key questions are: can these costs be controlled and who is going to pay?

But first, the resources have to be generated by a productive economy and a strong, stable system. The United States has been a world leader in the past century, in defense, technology, education, and generally in its influence on world matters. Whether it can keep this position in a global world is unclear.

It will depend on its ability to educate and take care of its people, innovate and create jobs, use its military force wisely, and preserve its democratic principles. Some of the risks to lose that prominence depend on endogenous forces – for example, the proper functioning of institutions – while others are exogenous factors, out of the US control, but which the country has to manage in a concerted and principled way.

Basic choices and values can allow the United States to avoid that decline. The migration of manufacturing out of the country has been originally motivated by lower labor costs elsewhere, and in some cases, looser regulations. As the world moves toward an economic equilibrium, this migration may revert itself. But the United States needs to seize opportunities to improve its ability to make manufacturing more efficient and bring it back home. Given the technological complexities involved in many such processes, the availability of high-skill workers is critical. It hinges upon the nature and the quality of US education, and also on its ability to retain top foreign workers, leaders, and students that contribute to its economic position.

It also hinges on protecting the US intellectual property. A major strength of the country is its entrepreneurial spirit, its ability to innovate and its capacity to fund the incubation and development of new firms, with big or small ideas. Some countries may develop competitive products independently. Others constantly try to steal the designs and innovation generated in the United States and sometimes market the products first. One way this happens is through the Internet.

Cybersecurity The Internet, the GPS, and other major cyber systems have brought capabilities unknown thus far, but brought with them new vulnerabilities. Cyberattacks are a relatively recent source of turbulences and concerns. Some involve routine hacking, others political or terrorist activities. Mostly, however, they come from organized crime (e.g., stealing of financial data) or state-sponsored organizations focused on US intellectual property, infrastructure, and defense systems. Given how much of the US economy depends on the Internet and the GPS, a well-targeted cyberattack can be devastating on the national scale. The worldwide nature of the Internet and social networks has thus fueled a new form of instability. The United States is still searching for cyber defenses and a surveillance system that balance individual rights and cybersecurity.

Managing the risks of cyberattacks is both an individual and a collective enterprise. It involves reducing both the chances of a successful attack and the consequences of a defense breach. Sharing information about attacks in the right forum, establishing legal boundaries for an appropriate response, and setting effective lines of defense are part of a global response. A challenge, both legal and practical, is whether and how to make hackers pay a price for their activities and to decide what kind of retaliation would be proportional and appropriate. The problem starts with tracking down the actual source of attack and responding in a measured manner. This is particularly critical against a determined, persistent state-sponsored aggressor if one wants to avoid an escalation of conflict. As with all risk management issues, priorities have to be set, and a balance has to be found between the frictions and costs of risk management measures and their effectiveness in protecting public and private assets.

Terrorism One of the most drastic changes in American life in the last decades has been the emergence of a new kind of terrorism, which took a new turn in September 2001 with coordinated attacks on the US territory. Since then, the multiple conflicts that affect the Middle East, the spread of Al Qaeda and the emergence of a so-called Islamic State have fueled a new wave of attacks and challenges to the security of Western states. These attacks are a manifestation of a deeper clash of cultures worldwide, rooted in religious, cultural, and political antagonisms that exploded in different forms and places, facilitated by a widespread use of instant communications. In the United States, these attacks have caused a deep sense of insecurity, and triggered a set of government measures that have kept the country relatively safe so far but have deeply affected people's life and psyche. Furthermore, these attacks have been linked to several military campaigns (notably in Iraq and Afghanistan) that have changed the face of war and emphasized the need for different types of weapons in a situation of asymmetric warfare. The economics of defense and military education are thus critical parts of budget choices that the United States has to make.

Given budget limitations, setting priorities among counterterrorism measures is essential since spreading resources evenly across the country and the different kinds of threats, while politically expedient would be clearly inefficient. Terrorist attacks can involve conventional explosives, dirty bombs (with nuclear material), bioattacks, and nuclear warheads. The challenge is to identify before the fact the most vulnerable and important targets and try to thwart the use of the most dangerous weapons. Picking up signals and anticipating terrorists' next steps is the role of both law enforcement and the intelligence community. In that instance, they have to work together. As massive amounts of information are gathered, one issue is to make sure that the privacy of individuals is respected to the greatest extent feasible. Another is that the signals be properly filtered so that the most relevant ones reach in time the decision level – along with the uncertainties that often remain.

One of the greatest risks of terrorist attacks is the explosion of a nuclear device. Facing that threat requires an organized response to the development of nuclear weapons, securing existing nuclear material including spent fuel, and containing nuclear proliferation. One can also argue, as have some of the authors in this book, that dismantling the US nuclear arsenal altogether will provide an example and a motivation for others to follow suit. Yet, given the ever-changing nature of world conflicts, the challenge for the United States is to find a balance between the risks of a nuclear accident and those of deterrence failure. The challenge starts with the control of nuclear material, which has to be generated at various levels of enrichment in the case of uranium, to be used in nuclear power reactors and medical devices, or to build weapons. Securing that material and discouraging the spread of high-level enrichment technologies is thus a security concern.

Energy, Environment, and Global Warming   The energy sector has been one of the most turbulent of the world economy, certainly so since the oil crisis of the seventies and the eighties. The cost of petroleum has fluctuated many times, and after skyrocketing in the last few years, has recently decreased, but for how long? The US dependence on oil imports has made it vulnerable not only to these price variations but also to political instabilities in producing countries. The increase of the production of shale oil in the United States has considerably reduced that vulnerability. At the same time, natural gas has emerged as a less expensive solution but environmental concerns regarding the fracturing of rocks ("fracking") remain and need to be addressed.

Burning hydrocarbons, oil, coal, but also to a lesser degree, natural gas, presents health and environmental issues associated with emissions of heat-trapping gasses such as carbon dioxide and other pollutants. Global warming is certainly one of the main turbulences on the world horizon. Given the stakes, efforts to curb it are imperative even if many uncertainties remain. One of them is the link between changes in the rate of CO_2 concentration in the atmosphere and the variation of worldwide temperatures. Furthermore, the consequences are uneven across the planet, as well as in the United States, where different regions will be affected unevenly by changes of precipitations, floods, and droughts, sea-level rise, and the frequency of related natural disasters such as hurricanes. The political challenge is to make now rational decisions under uncertainties about the costs and the benefits of curbing CO_2 emissions (how much reduction in temperature increases and in what time frame?). The question is crucial in the developing world, which needs large amounts of energy. Yet, the leadership has to come from developed countries that can afford that emission reduction in a rational and economic way and promote other forms of power generation. Nonpolluting renewables are an essential part of the solution, but so far entail considerable costs. New technologies need to emerge soon, among them energy storage on a large scale.

One solution is thus to build nuclear reactors, which do not pollute in nor-mal production times. There are currently 444 nuclear reactors and 63 reactors under construction worldwide, including 5 in the United States and many more are planned. But the nuclear power industry has been shaken by three major accidents since 1979 and the question of permanent storage of nuclear wastes needs to be resolved effectively and soon. Furthermore, in a highly reg-ulated environment, nuclear power may no longer be competitive with natural gas and its future in the United States is unclear at this time.

Part of the risk of worldwide reactors' construction is that of nuclear prolif-eration and the dual use of nuclear power generation. Plutonium and enriched uranium can be produced to make warheads and several countries may be constructing nuclear power plants to that end. Managing the risks of nuclear weapons proliferation – as those of cyberattacks – has to be part of an inter-national effort. Whether by setting an example or by other means, the United States alone cannot prevent the stealing and misuse of nuclear products. The problem needs to be addressed in concert with other nations and international entities through economic and diplomatic measures, in the realistic context of other nations' choices. Countries that feel threatened will develop weapons for their defense unless they believe that they will be otherwise protected.

Sustainable Urban Development Population moves in the last two cen-turies have triggered, in many places, the explosive growth of urban centers worldwide, often concentrated along the coastlines. Elsewhere, cities such as Detroit have shrunk when their industrial basis eroded at a time where their population was aging and resources were needed to support their pensions.

Some countries have managed urban growth rather successfully. Singa-pore, for instance, has grown into a stable, prosperous city in spite of huge challenges of dependence on neighboring countries for basic needs such as water. Other countries have struggled with an influx of new comers, immi-grants, and refugees, and the need to create jobs and to build and maintain sustainable infrastructures. In the United States, the challenge has been com-pounded by racial and social tensions, growing inequalities in the allocation of wealth and resources, and the requirements for cities and states to provide basic services through economic ups and downs. Environmental concerns in some regions have been coupled with threats of natural disasters such as tornadoes, earthquakes, droughts, floods, and tsunamis, requiring a clear and concerted approach in the design of structures and response capabilities.

Essential conditions for a sustainable urban development are also social stability and a sense of security. Growth has been stimulated in large part by automation and robotics but large segments of the population are being left behind. At this time, the United States is thus facing the risk of a social rupture due this income gap. In some regions – and as a consequence of extraordinary successes such as the Silicon Valley's – some cities are torn by the displacement of lower and middle classes with an influx of more affluent people. Public–private partnership is thus key to harmonious urban development, to encourage the construction of affordable housing, the creation of stable jobs and to avoid degradation of the social fabric and the detrimental effects of political polarization.

Some of the Trade-Offs

Time Horizon: How Long? One characteristic of the US system is an incentive to focus on the short term for at least two reasons. Industries and publically traded companies are rated on their immediate performance and stock value, with a focus on the next quarter(s). Yet, their long-term survival depends on their ability to anticipate changes of technology, the emergence of new demands and markets, and the behaviors of buyers and sellers. At the same time, the electoral calendar of politicians and the top people that they appoint sets strict time limits to their horizon for action. Urgent problems obviously need immediate attention, but long-term strategic thinking is essential to maintaining the US position.

The choice of a time horizon, both in public and private sector decisions, is obviously linked to the nature of the problems and to the shifts that can be anticipated. Policies related to long-term physical phenomena require a long-term view, with serious concerns for the well-being of future generations while respecting the needs of current ones. Yet, uncertainties about human evolution, innovation, adaptation, and desires limit the relevance of extremely long time horizons: thousands or millions of years may not make sense as a policy basis. More generally, for policies involving human behaviors, political systems, and economic situations, the pace of evolution of societies in essence limits the time horizon. What will be the world of our children and grandchildren is not entirely clear. Yet, their needs for housing, healthcare, or education will remain, and we must set the foundations to allow them to fulfill these needs.

Healthcare problems in particular, demand immediate attention since urgent current needs and demographics are known and the cost explosion unsustainable. Research, development, and adoption of new treatments require a longer lead time. The United States needs to find both a short-term and a long-term path to adapted care, oversight, and financial stability to provide innovative and safe medical care.

Similarly, in education, meeting the short-term needs of an increasing number of young people requires all reasonable attempts to improve students' performance now, and to close the gap between the poor and the privileged. Longer-term strategies include choices of the types of education required (e.g., college vs vocational education) and of education technologies. In the United States, this long-term view includes decisions not only about how to educate its own population but also foreign students who benefit from US resources, and often bring both a highly skilled labor force and connections to other countries.

In the energy sector, time horizon questions are particularly complex because of trade-offs and uncertainties in the pace of both the development of new technologies and the environmental effects of existing ones. Immediate energy developments or improvements are urgent in large parts of the world. In the United States, the energy needs include development of new technologies, investments in additional capacity, and improvements of the electric grid. Less clear are the time factors and the global costs of progress given the pollution from various methods of electricity generation and their long-term effects. These uncertainties include the time constants of global warming: some benefits of short-term reductions of CO_2 emissions are unquestionable but the delays of significant effects are less clear.

US Military Involvements   New trade-offs facing the US defense world are the results of simmering conflicts for example, with Russia, China and North Korea. Many new threats are affected by globalization, a change in the balance of power, and new forms of clashes fueled by economic, ethnic, and religious disparities. One of these trade-offs involves the need – and the duty, according to the UN charter – to protect populations at risks from murderous forces within their own country. A US intervention in these cases is costly and, in complex situations, the outcomes and the stability of the results are highly uncertain. Clearly, the United States cannot get involved in all internal conflicts and massacres. There too, choices have to be made and trade-offs have to be faced, reflecting a balance between duties, feasibility, and long-term effectiveness.

Private versus Public Funding   Given the costs involved in the different sectors discussed here, the country faces a fundamental trade-off between private and public funding. As shown by the authors of the different sections of this book, while a certain level of government support is needed, the private sector often brings additional competition and efficiencies.

In healthcare, the debate about funding revolves not only around questions of ethics and justice but also of incentives, effectiveness and, in the end, the best use by providers and patients of given amounts of money. In education, some argue against private funding because they believe that it perpetuates inequalities, while others think that freedom of enterprise and private investment is just the source of competition that can make all students better educated and more productive.

In the world of energy, public funds are needed to provide some of the infrastructure, but also to initiate the development of new technologies that may be critical when they are sufficiently mature. One question is how long these subsidies should last given the rates of progress and other demands on government resources.

In time of recession, Keynesian policies of state spending to prime the economic pump have shown great effectiveness in the past. They may have limited effect, however, when they have been used for too long and end up providing low returns and wrong incentives. More generally, the question of stimulus versus limitation of government debt and spending is at the heart of a debate that polarizes the US political and economic lives. Yet, there may be intermediate solutions. The basic questions are where do public sector investments show a return that justifies borrowing costs, and where may public/private cooperation be the optimal approach?

Privacy versus Intelligence Collection   Given terrorist threats, the United States faces a balance between individual freedom and privacy, and the need to collect information to thwart in time plans of attack and disruptions against the country. In addition, the intelligence community often faces uncertainties about the meaning of their observations, and the decision of when to issue a warning and how to describe what they know and what they do not.

One of the major problems of privacy versus collection lies in the global history of intelligence and surveillance, which have been used at times, to threaten forces of change that disturbed the powers in place. The US intelligence community in that respect is in a tough spot. Its technological capabilities allow collection of huge amounts of information. Storing it indiscriminately in vast quantities to find networks where they may exist is an obvious but at times disturbing solution. In addition, to protect its sources and methods, the IC cannot disclose its successes, but its failures are often clear and easy to point to after the fact. Suspicions of collaboration with people in power for their benefits and of spying among allies are widespread, both nationally and internationally. Yet, international collaboration is essential. The problem of freedom versus security is a fundamental constitutional one. Yet, it does not have an obvious solution and needs to be addressed upfront as a trade-off in times of turbulences.

Collaboration versus Competition with Other Countries   In economics as well as defense, there are opportunities for collaboration with entities and countries that have traditionally been US competitors or opponents in one aspect or another. In these cases, diplomatic and economic choices reflect a delicate balance between collaboration and competition, protecting the US intellectual property while taking advantage of world markets, or finding alliances when needed to address problems such as nuclear proliferation. "Coopetition" is a term that has been applied to these situations. It requires flexibility and a strategic view, as well as a realistic perception of immediate problems and changes that may come; in other terms: "trust and verify."

Environmental Concerns versus Economic Realities Urban centers are magnets when their economies thrive. Population movements within the country reflect these opportunities. But that flux of population, which follows the ups and downs of different economic sectors, sometimes threatens the ecology of the regions involved (e.g., oil in Alaska or North Dakota). In these cases, environmental concerns have to be balanced against population employment and needs, including the production of food and energy at affordable prices. Innovation in the quest for that balance is key to public health, risk management, and affordable living standards. But it also requires that the public be educated to understand the relative magnitudes of the various risks, and the effectiveness of the different solutions proposed. The communication of risk in particular must be free of undue pressures from the industries whose interests are at stake. It also must also represent objectively the tradeoffs involved, as some groups may want to scare the public to achieve political objectives. At the receiving end, understanding the message still requires a basic level of education and numeracy to allow comparison of solutions. Environmental problems are too important to be addressed without a basic understanding of what is known and what is not about the fundamental mechanisms involved in the ecological balance.

As demonstrated in this book, these trade-offs are viewed differently by opposing parties with different political traditions and personal preferences. The objective here is to present a few examples and recommendations from various reasonable perspectives, sometimes leading to opposite conclusions.

A Proactive Approach to Risk Management

Times of turbulences and instabilities when instant communications have made the fuses very short require even more vigilance as risks can escalate into conflicts and disasters. Uncertainties have always been complex aspects of policy and strategy decisions. In the end, they are based on how much is known – or imagined – and what risk one is willing to take. Uncertainties, of course, cut both ways with prospects of a better and longer life, along with possibilities of decline, conflicts, and losses. Given resource and time limitations, investments have to be prioritized to minimize the risks faced by the nation, its industries, and its people, while providing jobs and fulfilling basic needs.

This is why quantifying the risks can be helpful if not easy. Instead of waiting for a system to collapse, recognition, assessment, and prioritization of risks – big and small – are essential to an effective, proactive management. This requires close monitoring and recognition of precursors and near misses, a challenge when the signal is hard to separate from the noise. But in spite of screaming headlines, media attention, and anxieties, the most threatening risks may not be those of extremely rare events. The terms "black swans" and "perfect storms" are often overused – especially after the fact – to describe very unlikely situations which, in reality, could and should have been anticipated given the signals. Still subtler may be the anticipation and detection of "brewing bubbles," these small incidents and perturbations – social, political, economic, or environmental – that may coalesce and result in massive disruptions.

In all these cases, the lead time provided by signal observations is critical to an effective response, provided that there is the will, incentives, and resources to face the challenges. This is a question of leadership to be addressed at the top levels of organizations.

OPPORTUNITIES

In spite of turbulences and risks, the United States is in a unique position to exploit the opportunities that are the flip side of these challenges thanks to its flexibility and ability to quickly face and manage transitions.

Creativity and Innovation

So far, almost nowhere in the world is there the level of innovation and creation that the United States has been able to develop and maintain on a large scale. Many countries, such as South Korea, have been innovative and entrepreneurial in specific sectors. Yet, very few have been able to create, through education and funding, an environment that fosters the creation of enterprises worldwide, across a large economic spectrum. To keep this spirit alive, the United States needs to maintain or to restore its quality of life and education, its tolerance for failure, the inclusion of immigrants at all levels of income and education, the availability of funds for startups, and a culture of entrepreneurship supported by venture capital and management guidance. As in the past, the companies that are now thriving may come and go. Other successful ones will replace them as long as the United States stays at the cutting edge of both technological and social development.

Energy Independence

The oil market has been a major source of turbulence worldwide, but in recent years, the exploitations of shale oil and natural gas have provided an opportunity for cheaper energy. Until wind and solar energy become truly competitive, this decrease in energy costs offers economic opportunities to both producers and consumers. And there are several alternatives to the current solutions.

Carbon capture, although costly at this time, may be a possibility if affordable solutions are found. Electric cars are a valuable option to reduce emissions from conventional engines as long as electric production is cheap and clean enough to meet economic and environmental goals.

Nuclear energy is another option, which at this time, is adopted by many countries worldwide. To be welcome in the United States, it has to be considered safe and economical compared with alternatives. Its adoption implies that the reactors are properly constructed and operated, that the United States solve the problem of long-term storage of nuclear wastes, and that controls are in place to provide protection against the misuse of spent fuel. Small modular nuclear reactors may be part of the answer as they are safer, smaller, and generate less waste. Meanwhile, a 1975 law in the United States forbids the reprocessing of spent fuel, even though a large fraction of the energy is still present in the wastes, which cannot be endlessly stored on-site. Therefore, there is no perfect, long-term solution that can be implemented at this time.

In the long term, the availability of new, renewable sources of energy will provide some stability. Indeed, the promises of solar and wind energies are immense. The challenges are still high cost and dependency on the weather and climate in the absence of large-scale effective storage in areas where sunshine is rare and winds uneven.

Reduction of the US Nuclear Weapons Stockpile

Given the changes in the world of defense and the costs of maintaining a large nuclear weapons stockpile, the United States has an opportunity to decrease considerably the size of its nuclear arsenal, while possibly upgrading it. This would reduce the risks of a technical failure and an accidental attack. Plus, some argue that nuclear weapons are ineffective, anyway, against a non-state adversary and therefore that a much smaller arsenal – if not a zero one – is sufficient. Meanwhile, the United States is facing considerable challenges since major powers such as Russia and China have chosen to maintain and modernize their nuclear arsenal at the same time as they try to expand, and minor powers like North Korea have become true nuclear threats. The US reduction of its nuclear stockpile may thus be limited by the need for an appropriate deterrence capability but there is certainly an opportunity to reduce the costs and the risks that have built up during the cold war and remain a nightmare for those involved in the dilemmas of that era.

Addressing the Size of the Defense Budget

More generally, the structure of the US defense system and the new kinds of conflicts that the country is facing present an opportunity to plan for a different kind of defense structure and perhaps, a reduction of the US defense budget. To be sure, emerging situations in Eastern Europe, the Middle East, and South East Asia have eroded the "peace dividends" of the end of the cold war. Asymmetric conflicts may require more special forces although they cannot be expected to replace conventional ones. New types of ships and aircraft, manned and unmanned, may provide more flexibility across services, and the improvement of the quality of training and education of US officers and enlisted personnel has improved their effectiveness. Altogether, and although opinions are divided on this point, there may be an opportunity, in times of budget reductions, to enhance both defense capabilities and effectiveness.

Teaching on a World Scale

The quality of education in the United States needs to be maintained and improved. There is a new opportunity of doing so, both at the national and international scales, with the development of Internet-based teaching through massive online courses. Such courses, obviously, cannot provide the one-on-one experience that students can get in a regular classroom. To permit the widespread implementation of these courses, many problems need to be addressed, from the protection of intellectual property to the very quality of teaching online. But they provide an economic alternative to a higher-education system that has become too expensive for many, at a time when knowledge and education are critical to employment and development.

Social Media

Instant communication has drastically modified the lives of people at the individual and collective levels. Possibilities of contacting friends on the spot or setting up new business or political connections have modified drastically the nature of relations. Checking online the availability and prices of goods and services allows more efficient use of individual resources. Instant communications of news across the world have flooded people with information – true or not – and facilitated the coalescence of insurgent forces, while providing an opportunity to bring help and support where needed and a diversity of information sources and viewpoints. The opportunities provided by social media are both extraordinary and intractable. Yet, a new awareness of global problems brings the perspectives needed to address these challenges on a global scale with a better sense of the consequences of individual and national decisions.

Population Migrations

Worldwide awareness of opportunities elsewhere and war situations and crises has set waves of massive migrations, often from the South to the North and more recently, from the Middle East with the hopes and dreams of a better and safer life. Immigration in the United States is a center of debate. What is clear is that lower-skilled manpower on the one hand and new intellectual and leadership resources on the other present an opportunity to the country, now as it has in the past. From the agricultural workers to the high-tech entrepreneurs, immigrants bring the human resources indispensable to the continuing competitiveness of the United States. The challenge, as always, is to control the immigration flow and integrate the new comers, with their values, their customs, their traditions, and their needs. The United States can do it but has to be willing and able to provide the security, the education, and the healthcare that are required for its new citizens to prosper. The country also needs to manage in a sustainable and harmonious way the urban growth that comes both with immigration and internal migrations.

Globalization

Globalization is a source of both turbulences and opportunities. It permits a movement of resources, people, and capital unknown so far, with the benefits of freedom and efficiencies. It is also creating a world of international competition that profoundly affects some countries' domestic life and, in some cases, globalization deepens the gap between richer and poorer regions. Instant market response and population movements thus cause local problems in some areas, while facilitating access to cheaper goods and services. Global economic stability given political and cultural disparities will take time to settle – if it ever happens – and other problems are bound to emerge. In this globalization context, the United States has to remain competitive, especially with countries where lower labor costs prevail. This requires domestic policies – fiscal, monetary, educational, and so on – that promote growth, innovation, and entrepreneurship, all critical to US competitiveness, employment, and wellbeing. It also requires strengthening the institutions, embracing diversity while sticking to fundamental values, and supporting a tradition of technological innovations, social support, and substantial investments in new enterprises.

Obviously, the domains that we chose to address in this book do not represent an exhaustive list of the challenges that the United States and other countries face; neither do the aspects of these domains that our contributors chose to discuss. It is not our goal to present definitive solutions to the problems that are discussed. The hope is that this book can provide useful information and a diversity of perspectives, shed some light on some policy questions, and help bridge political gaps toward effective solutions through a truly democratic process.

OVERVIEW OF CONTRIBUTION

Norman R. Augustine, former CEO of Lockheed Martin, sets the stage by asking "Can America Still Compete?" He argues that a strong economy requires abilities to compete globally, noting that a majority of GDP growth in the United States is due to advances in science and engineering. However, the education system is failing to produce the needed scientists and engineers. The problem, he observes, is how we spend our K-12 resources. He asks whether we can continue to attract talent from overseas and suggests that this practice, at least, in part, drives decisions to put R&D facilities overseas. He suggests a variety of initiatives to remediate these problems and enhance competitiveness.

CAN AMERICA STILL COMPETE?

Norman R. Augustine

None of the challenges America confronts today can be met without a strong economy; be it providing healthcare, assuring national security, supplying energy, preserving the environment, or enhancing the overall quality of life of its citizenry. But, in order to create a strong economy – and thus the resources the government will need to address these issues – will demand that the nation's citizens be able to compete globally for jobs. By my own calculation, for many decades each percentage point of GDP growth has been accompanied by approximately a 0.6% point growth in jobs.

Furthermore, a number of studies, one of which was the foundation for a Nobel Prize, have demonstrated that 50–85% of the growth in GDP in recent decades is attributable to advancements in just two fields: science and engineering. Unfortunately, our nation today is not competitive at producing scientists and engineers. There are, of course, those who argue that we already have too many scientists and engineers – which is indeed true if we elect to continue our policy of underinvesting in science, engineering and the future. But if we wish to create jobs for the workforce *as a whole*, we must invest more in producing world-caliber scientists and engineers.

But in this regard, by global standards, America's educational system is failing. To begin with, 15-year-olds in our public schools rank 22nd in science and 25th in math among the 34 nations participating in international tests. It is, therefore, perhaps not surprising that our universities now produce

Perspectives on Complex Global Challenges: Education, Energy, Healthcare, Security and Resilience,
First Edition. Edited by Elisabeth Paté-Cornell, William B. Rouse, and Charles M. Vest.
© 2016 John Wiley & Sons, Inc. Published 2016 by John Wiley & Sons, Inc.

relatively few scientists and engineers. In a recent survey of the fraction of degrees at the baccalaureate level that are awarded in engineering, the United States finished 79th out of the 93 nations examined. The only countries behind the United States are Bangladesh, Brunei, Burundi, Cambodia, Cameroon, Cuba, Zambia, Guyana, Lesotho, Luxembourg, Madagascar, Namibia, Saudi Arabia, and Swaziland.

It should be noted that the problem in US K-12 education is not the lack of funds: our nation spends more per student than any other nation on the planet except one. The problem is *how* we spend the funds.

For many years, America has managed to be competitive by attracting to our shores the best and brightest from abroad – a source that is still providing about two-thirds of the engineering PhD recipients at America's own universities. But this is a dependency that is becoming unsustainable due to the growth of opportunities abroad and disincentives here.

I have participated in over 500 board meetings of Fortune 100 companies and watched America's firms discover that the solution to their talent problems is straightforward: simply move factories and R&D centers and administrative functions overseas. In fact, when global firms now decide where to build new facilities, 77% of the time the answer is *not* in the United States. For example, Microsoft has said it was establishing a software development center employing 800 engineers just across the border in Canada because of the barriers to attracting talent to the United States. Steve Jobs told the President of the United States that the reason his firm employed 700,000 people abroad was because they cannot find 30,000 engineers here. And Intel's Howard High stated that, "We go where the smart people are. Now our business operations are two-thirds in the US and one-third overseas. But that ratio will flip over in the next ten years." Or, in the words of DuPont's then-CEO, Chad Holliday, "If the US doesn't get its act together, DuPont is going to go to the countries that do."

What then might America do to be competitive in the global economy? For openers, we could bring the Free Enterprise System – that has been so successful in higher education and other fields – to K-12 education. This involves paying a physics teacher more than a phys-ed teacher; paying a great physics teacher far more than a good physics teacher, and helping a poor physics teacher find a more suitable career. In our universities, it includes rebalancing the rewards for extraordinary faculty performance with the rewards for successful football coaching.

Furthermore, as a jump-start, the federal government could each year fund 10,000 competitively awarded scholarships to US citizens to attend US universities and study math, science, or engineering – with the understanding that upon graduation they will teach in a public school for at least 5 years.

The continued excellence of our higher education system, as well as of our economy, could be enhanced by greatly increasing our investment in basic research; by having states reject the currently widespread and damaging policy of disinvesting in, and thus quasi-privatizing, our great public universities; and by controlling escalating tuition by selectively adopting technology in teaching. Our universities, which have barely changed their pedagogy in over a century, have the opportunity to become much more efficient through distance learning, innovative classroom practices, computer tutoring, and more.

The federal government could eliminate wasteful regulations that consume research dollars. It could also reduce to 9% taxes on repatriating the $2.1 trillion that US industry has sitting abroad to avoid double taxation, and use the proceeds to establish an endowment for an independent research funding foundation. The loopholes in the tax structure that distort investment practices and favor only the wealthy could be eliminated. We could change the tax structure such that capital gains on assets held less than 1 day would be taxed at a 99% rate, and those held more than 10 years at a 1% rate – adopting whatever monotonic curve connecting these points is necessary to produce the desired federal revenue. Such a policy would bring an end to the current distortions in business practices caused by "day traders" and perhaps even give CEOs the courage to invest for the long term.

But to bring about such dramatic changes will require breaking the partisan deadlock that has paralyzed the nation's capital. This might be accomplished by establishing, outside government, a modest-sized but permanent institution, governed by a board consisting of highly respected, retired individuals from both political parties as well as independents, to actually write potential legislation that addresses the most critical and confounding problems facing the nation. Congress would then be required to vote on that legislation, "up or down," with no amendments permitted.

America's structural problems are such that the Band-Aid solutions of recent decades are no longer viable; major surgery is in order.

SECTION I

EDUCATION

1

INTRODUCTION

According to *The Economist* (2014), the United States ranks 14th in educational performance among developed countries in 2014, far behind the top five – South Korea, Japan, Singapore, Hong Kong, and Finland. This is an improvement from being 16th ranked in 2012, but a far cry from being ranked 1st two decades ago. Put simply, other countries have been investing and innovating while the United States has lagged in these areas.

More specifically, the US high school graduation rate ranks 21st, much lower than the top five – Portugal, Slovenia, Finland, Japan, and the United Kingdom. The United States is ranked 48th in quality of math and science education. Overall, the United States no longer leads education and innovation in science, technology, engineering, and math. For example, Europe and Asia-Pacific now produce two thirds of published research papers.

Thought leaders in the US educational system saw this coming. Reports from blue ribbon commissions included *A Nation at Risk* (National Commission, 1983) and *Before It's Too Late* (National Commission, 1999). More recently (National Academies, 2007), *Rising Above The Gathering Storm* outlined the issues and approaches to addressing them. This section of the book provides a range of perspectives on the central issues for both K-12 and university education.

Throughout the many debates on K-12 education, there was usually great confidence that the university system in the United States was in fine shape.

Perspectives on Complex Global Challenges: Education, Energy, Healthcare, Security and Resilience,
First Edition. Edited by Elisabeth Paté-Cornell, William B. Rouse, and Charles M. Vest.
© 2016 John Wiley & Sons, Inc. Published 2016 by John Wiley & Sons, Inc.

This system has benefited from strong leadership over the past half century, as briefly outlined later. However, the system has recently come to face significant challenges as discussed later in this introduction.

The pre–World War II 20th century provided a hotbed of research in physical sciences and mathematics. Physics and computing are of particular note. However, the modern research university, particularly in the United States, emerged following World War II. Vannevar Bush, a 20th-century leader in engineering and science, was instrumental in defining the vision.

Bush articulated the central principles in *Science: The Endless Frontier* (1945):

- The federal government shoulders the principal responsibility for the financial support of basic scientific research.
- Universities – rather than government laboratories, nonteaching research institutes, or private industry – are the primary institutions in which this government-funded research is undertaken.
- Although the federal budgetary process determines the total amount available to support research in various fields of science, most funds are allocated not according to commercial or political considerations but through an intensely competitive process of review conducted by independent scientific experts who judge the quality of proposals according to their scientific merits alone.

Perhaps not surprisingly, Bush's home university, the Massachusetts Institute of Technology (MIT), was very successful in adopting these principles. James Killian, MIT president from 1949 to 1959, notes that "From MIT's founding, the central mission had been to work with things and ideas that were immediately useful and in the public interest. This commitment was reinforced by the fact that many faculty members had during the war direct and personal experience in public services" (Killian, 1985, p. 399).

He reports that MIT's relationship with the federal government reached new heights with World War II:

- MIT took on critical challenges, for example, the Sage missile defense system and the Whirlwind computing project.
- Faculty and alumni serving in important advisory roles in the federal government.
- Faculty, including two MIT presidents, served in senior executive positions, on leave from MIT.

As a consequence, MIT became and remains a national resource, perhaps the key player in "big science." In the process, MIT was transformed into a university. This was facilitated by several factors (Killian, 1985):

- A single, unfragmented faculty in consort with one central administration.
- Close articulation of research and teaching, of basic science and applied science.
- Continuous spectrum of undergraduate and graduate studies.
- Mobility of ideas resulting from the high permeability of the boundaries of both departments and centers.
- The extensive interconnection of its buildings.

MIT, and a handful of other leading institutions such as the University of California at Berkeley, California Institute of Technology, University of Illinois, and Stanford University, led the way defining the nature and "rules of the game" for research universities. In the process, science and technology has become central to our economy. As Richard Levin, former president of Yale University, indicates, "Competitive advantage based on the innovative application of new scientific knowledge – this has been the key to American economic success for at least the past quarter century." (p. 88). He asserts that the success of this system is evident: The United States accounts for 33% of all scientific publications, has won 60% of Nobel Prizes, and its universities account for 73% of papers cited in US patents (Levin, 2003).

However, it is not clear that this traditional model is sustainable. James Duderstadt (2000), former president of the University of Michigan, summarizes several areas of concern identified in a National Science Foundation study:

- Public support has eroded with continual decline throughout the1990s.
- Limits on indirect costs have resulted in cost shifting.
- The focus on research funding has changed the role of the faculty.
- Increased specialization has changed the intellectual makeup of academia.

He argues that the real issue is a shifting paradigm for universities. National priorities have changed, although recent security concerns have moderated this trend. The disciplines have been deified, yielding a dominance of reductionism. This presents a challenge for interdisciplinary scholarship, particularly in terms of valuing a diversity of approaches and more flexible visions of faculty career paths. At the same time, undergraduate education is receiving increased attention, as have cultural considerations that, he cautions, tend to encourage "belongers" rather than "doers."

Ruminating on the roles of publicly supported research universities, Duderstadt suggests several possibilities for strategies that universities can pursue as a response to current challenges:

- Isolation: Stick with prestige and prosperity, for example, MIT, Caltech, Princeton, Chicago.
- Pathfinders: Participate in experiments creating possible futures for higher education.
- Alliances: Ally with other types of educational institutions.
- Core-in-Cloud Models: Elite education and basic research departments surrounded by broader array of entities.

Derek Bok, former president of Harvard University, addresses the future of universities in light of many recent trends (Bok, 2003). He is particularly concerned with the commercialization of the university in response to a plethora of "business opportunities" for universities. He notes "Increasingly, success in university administration came to mean being more resourceful than one's competitors in finding funds to achieve new goals. Enterprising leaders seeking to improve their institution felt impelled to take full advantage of any legitimate opportunities that the commercial world had to offer" (p. 15). He argues that this increased focus on commercialization may jeopardize the focus on education and learning.

Bok recognizes that this shift is nevertheless taking place. He cautions, however, that universities typically face several challenges that can hinder entrepreneurial aspirations. Bok summarizes these challenges, "On three important counts, the environment in most research universities does not do enough to encourage the behaviors needed for the sake of the students, the society, and the well-being of the institution itself" (pp. 23–24).

- Efficiency: "University administrators do not have as strong incentives as most business executives to lower costs and achieve greater efficiency" (p. 24).
- Improvement: "A second important lesson universities can learn from business is the value of striving continuously to improve the quality of what they do" (p. 25).

- Incentives: "Left to itself, the contemporary research university does not contain sufficient incentives to elicit all the behaviors that society has a right to expect" (p. 28).

These seem like reasonable challenges, at least for businesses. However, Bok argues "Leading a university is also a much more uncertain and ambiguous enterprise than managing a company because the market for higher education lacks tangible measurable goals by which to measure success" (p. 30). Furthermore, he asserts "Presidents and deans are ultimately responsible for upholding basic academic values but they are exposed to strong conflicting pressures that make it hard for them to carry out this duty effectively" (p. 185).

We would expect that market forces would resolve these pressures. However, Bok reasons "Neither the profit motive nor the traditional methods of the research university guarantee that faculties will make a serious, sustained effort to improve their methods of instruction and enhance the quality of learning on their campuses" (p. 179). In other words, we cannot expect an organically based transformation of academia, despite financial and social forces for fundamental changes. There is a fundamental tension between what is naturally happening in research institutions (i.e., increased focus on the external viability of research), and the way in which this is being managed, or not managed, within the same universities. The lack of attention and process in the midst of this evolution could halt the progress and risk the outcomes of the changes.

Charles Vest, former president of MIT and president of the National Academy of Engineering, provides both retrospective and prospective views of academia (Vest, 2007). He reviews the last half century in terms of how federal support has shaped the education system. He considers how the private sector, via philanthropy has shaped academia. The impact of September 11, 2001 on the openness of the research enterprise is his third topic. Finally, Vest considers how the Internet will likely impact the higher education enterprise. Perhaps not surprisingly, his projections and the actual consequences have been profound.

The Great Recession of 2007–2009 accelerated the decrease of state support for public education. Between 1980 and 2011, state support decreased by 15–70%, depending on the state. As a result, tuitions and fees have soared at public institutions, funded by government-backed student loans, and leading to student debt exceeding all credit card debt. In parallel, costs of operations at universities have soared, with most of the increases coming from administrative costs rather than delivery costs, for example, faculty salaries. Consequently, higher education has become the poster child for runaway costs, replacing healthcare that now seems much more controllable.

This has led to new prescriptions such as Christenson and Eyring (2011) who explore how universities can find innovative, less costly ways of performing their uniquely valuable functions and thereby save themselves from decline. The authors outline the history of Harvard University and how various aspects of academia were defined by Harvard leadership in response to issues and opportunities of the times. They explore the strategic choices and alternative ways in which traditional universities can change to ensure their ongoing economic vitality. They emphasize the need for universities to address key tradeoffs and make essential choices as they decide how to compete.

DeMillo (2011) addresses the challenges faced by "the Middle," the 2,000 universities that are not part of the "Elite" – those with $1 billion plus endowments – and also face stiff competition from for-profit online universities. Computer-based and online technologies, such as MOOCs, and new student-centric business models are discussed. The book culminates in 10 rules for 21st-century universities, expressed in terms of defining value and becoming an architect of how this value is delivered.

It is clear that a transformation of higher education will inevitably happen. While the elite schools may sustain their current business models, the vast majority of educational institutions will not. Creative destruction will be rampant in education. The transition will likely be quite painful, but the end result promises much better outcomes at much more reasonable costs.

OVERVIEW OF CONTRIBUTIONS

As discussed earlier, in the last decades, the quality of K-12 education in the United States has fallen in worldwide rankings. The costs of tertiary education are growing at a higher rate than inflation and the gap between rich and poor students' performances is widening. On a brighter side, technologies such as computer-based learning, and now, massive Internet courses can be parts of the solution. But nothing will replace good teachers. What is at stake is not only US competitiveness but also national security.

Craig R. Barrett, a former professor of engineering and CEO of Intel, identifies three keys to success in "K-12 Education Reform in the United States." The three keys are quality of the teachers, high expectations, and competition for better performance in the delivery of knowledge. A strong advocate of competition in the world of education, he believes that charter schools are part of the solution, along with the option of public schools and homeschooling. Students should be motivated to believe that what they learn is relevant to their future, and parents should be free to choose the system in which their children are educated. Their choice should be independent of their financial means, and supported by an annual allocation of public education funds that they should be free to spend elsewhere. Focusing on the quality of education, he argues that teachers can be helped but not replaced by intelligent use of technology in the classroom, and that they should first be experts in the field they teach, then in methods and practices of pedagogy. He also recommends worldwide benchmarking of tests and curricula. He concludes that these goals – and especially competition in the world of education – will be reached only with the political will and support of leaders and voters.

Deborah J. Stipek, former dean of the Stanford University School of Education, argues that we should "Secure America's Economic Future By Investing in Young Children." She emphasizes the economic value of early education and the need for investments in preschools. She argues that the benefits of these investments go well beyond their economic return (in a recent study, 18% for a program in Chicago), and that they bring a host of other social benefits. Among others, they reduce the need for special education, and lower the rates of teenage pregnancy and of incarceration. They improve future learning abilities by providing the neurological foundations of early development. Later in age, she argues that one effect of preschool education is to bring more stability in students' lives and an opportunity for children from low-income families to catch up to some extent, with their wealthier peers. She concludes that such a long-term investment will take political courage, but will be critical in the future competitiveness of the country.

Linda Darling-Hammond, a professor of education at Stanford University, addresses "The Future of Teaching in the United States." She focuses on the social disparities that the US education system creates and perpetuates. This is true not only among American students from different social classes, but also between US students and those from other countries of the OECD. To bridge that gap, she advocates a centralized, state-run system, with adequate, equitable resource allocation. The objective is to avoid the disparities created by different levels of local funding between poor and wealthy communities. She also emphasizes the importance of competitive selection, appropriate compensation and evaluation of teachers. In practice, she recommends continuing improvement of teachers' education, a licensing system based on performance, regular and meaningful teachers' evaluations, and career ladders that motivate them by offering opportunities for advancements.

Lawrence S. Bacow has had considerable experience in university management, at MIT, Tufts, and Harvard. He addresses "The Conundrum of Controlling College Costs," deploring the rise of college and university tuitions beyond the rate of inflation and points to the systemic contradictions that have led to this situation. Students and parents demand more offerings and student services, and the competition for star teachers, at ever-increasing costs, is intense. At the same time, the decline of state support in the current economic climate contributes to the gap between costs and revenues. He sees an opportunity in the use of technologies and massive online teaching. But mostly, he argues that collaboration among academic institutions might be the only way to control these costs, and that this can be achieved, if needed by modification of the legal system, without collusion and breach of antitrust laws.

William J. Perry, a former Secretary of Defense and an emeritus professor of engineering, discusses "Military Education." He describes the massive improvements that have taken place in military education since World War II. From a low point in morale at the end of the Vietnam war, he sees the rise of a new generation of leaders – commissioned and noncommissioned officers – educated in top institutions both civilian and military. He believes that this improvement in higher education among the leaders of the armed services has contributed to the strength of US military forces that have never been as well prepared for their mission. He points out in particular, the role of understanding cultures and languages in the type of conflicts in which the United States is engaged at this time and the benefits of training foreign officers in US military institutions. As an example of how to achieve a high level of knowledge without an increase of costs that the country cannot afford, he describes a computer-based tutoring system that has been helpful to the military. While this is not a new idea in itself, its actual, successful implementation shows the potential of this approach to basic education when here can be little direct contact between teachers and students.

REFERENCES

Bok D. *Universities in the Marketplace: The Commercialization of Higher Education.* Princeton, NJ: Princeton University Press; 2003.

Bush V. *Science – The Endless Frontier: A Report to the President on a Program for Postwar Scientific Research.* Washington, DC: National Science Foundation; 1945.

Christenson CM, Eyring HJ. *The Innovative University: Changing the DNA of Higher Education from Inside Out.* San Francisco: Jossey-Bass; 2011.

DeMillo RA. *Abelard to Apple: The Fate of American Colleges and Universities.* Cambridge, MA: MIT Press; 2011.

Duderstadt JJ. *A University for the 21st Century.* Ann Arbor, MI: University of Michigan Press; 2000.

Economist. 2014. The learning curve. Available at http://thelearningcurve.pearson.com/2014-report-summary/. Accessed 2016 Jan 19.

Killian JR Jr. *The Education of a College President: A Memoir.* Cambridge, MA: MIT Press; 1985.

Levin RC. *The Work of the University.* New Haven, CT: Yale University Press; 2003.

National Academies. *Rising Above the Gathering Storm: Energizing and Employing America for a Brighter Economic Future.* Washington, DC: National Academy Press; 2007.

National Commission. *A Nation at Risk: The Imperative for Educational Reform: Report from the National Commission on Excellence in Education.* Washington, DC: US Department of Education; 1983.

National Commission. *Before It's Too Late: A Report to the Nation from the National Commission on Mathematics and Science Teaching for the 21st Century.* Washington, DC: National Academy Press; 1999.

Vest CM. *The American Research University from World War II to World Wide Web.* Berkeley, CA: University of California Press; 2007.

2

K-12 EDUCATION REFORM IN THE UNITED STATES

Craig R. Barrett

In the middle of the 20th century, the public K-12 education system in the United States was considered one of the best in the world. Even though the graduation rate was only about 70%, follow-on postsecondary degree attainment was the highest in the industrialized world, giving the United States a very competitive work force. Despite this success, there were warning signs that the system was in trouble.

Shortly after Sputnik was launched in 1958, a K-12 study entitled "In Pursuit of Excellence" was published that detailed needed improvements in K-12 education. Following that report, a new study has been issued every decade or so repeating the same suggestions for improvement. These include A Nation At Risk (1983), Before It's Too Late (1999), Rising Above The Gathering Storm (2005), and a study by the Council on Foreign Relations relating deficiencies in K-12 education to national security (2012). In addition, several studies were carried out by the National Governors Association that suggested needed improvements.

Despite all these studies, the performance of the K-12 system has remained largely unchanged. In over 40 years, the scoring of 17-year-olds on basic international benchmarked tests in language arts, mathematics, and science has remained flat. The assessments done 50 years ago indicating that students

Perspectives on Complex Global Challenges: Education, Energy, Healthcare, Security and Resilience,
First Edition. Edited by Elisabeth Paté-Cornell, William B. Rouse, and Charles M. Vest.
© 2016 John Wiley & Sons, Inc. Published 2016 by John Wiley & Sons, Inc.

had great difficulty reading for comprehension, constructing oral or written arguments, solving multistep mathematical problems, and understanding science, geography, and civics are still the same today. What was once a leading system in the world is now mediocre by international standards with US children ranked in the lower half of the industrialized world in math and science and only about average in language arts (English). While the rest of the industrialized world has shown improvement, the United States has stayed flat in performance. And a follow-on result is that the United States, which used to have the highest postsecondary degree attainment in the world, is now ranked about 13th. These educational challenges are part of the reason the United States has recently dropped in World Economic Forum competitiveness rankings from #1 to #7.

Numerous studies of high-performing education systems throughout the world have all concluded that there are three common features for successful systems. First, the quality of the system is directly dependent on the quality of the teachers. While in most high-performing systems teachers are drawn from the top tier of university graduates and are accredited in the subjects they teach, studies in the United States show quite the opposite. Graduates of schools of education in the United States typically represent the lower third of college entrants in terms of academic performance and many math and science teachers in the United States are not accredited in the subjects they teach. As teachers in the public K-12 system are generally required to have a teaching credential granted from a school of education, the system biases against top performing university graduates unless those students specifically choose to enroll in a school of education and compete the education degree requirements. In addition, once in the teaching system, K-12 teachers in the United States are hardly treated as professionals as they are given little time for professional development, are usually not graded on the performance of their students, and are typically granted tenure after only a few years on the job.

The second factor key to the success of any education system is the high expectation level placed on both curricula and tests. There are many examples of internationally benchmarked exams given around the world (NAEP, PISA, Cambridge, etc.) and many high-performing countries base their classroom study and tests on these measures of success. In the United States, there have been two reasons why this has not been the case. First education is locally controlled in the United States with more than 15,000 locally controlled school districts. Each of these districts controls the curriculum and often it is set at the lowest common denominator to accommodate the slower learning students. In addition, measures of success are often comparisons with other local school districts and not with international standards of performance. As a result, US children compare poorly on the international scale of success with only about one-third being ranked proficient by NAEP standards. Attempts to upgrade curriculum standards have often failed as social pressure for grade promotion trumps higher expectations of performance.

The third characteristic of high-performing education systems is something called tension. Tension comes in various forms and can generally be described as pressure or competition from within for better performance. This can occur via the use of internationally benchmarked tests to compare local performance to the best in the world. Or it can come from local competition from non-traditional public schools (charter schools) or private schools that break the monopoly of the standard public school system and give children and parents a choice in which school to attend. And it can come from performance grading of teachers, administrators, schools, and school districts where promotions, tenure, salaries, and funding are driven by performance measures rather than time in seat metrics. Whatever the cause of the tension, its presence promotes improved performance and provides a baseline for continuous growth in the system. While the United States is still experimenting with tension in the form of limited competition from charter schools and grading of schools, teachers, and administrators, the bulk of the US K-12 education system is still a monopoly void of competitive pressure.

Building upon the three factors key for the success of any K-12 education system, there are five things that might be done to improve the US system. All of these five suggestions have been made many times before without impact, but perhaps the reasons why they might be accepted today are simply the increased awareness of the deficiencies of the US system and the increased demand for a well-educated workforce to compete in the 21st century.

GREAT TEACHERS

Schools of education in the United States have overly focused on the pedagogy of teaching and not the content of teaching. Great teachers need to know content as well as how to teach. There are many examples of school systems (charter schools and private schools) where it has been shown that content experts in the classroom excel over pedagogy experts. Increasingly schools of education are moving away from a strict pedagogy approach and requiring content expertise of teachers. The UTeach program introduced at University of Texas is a prime example of the new model for schools of education. Demand that prospective teachers be content experts first and foremost, with adequate skills in how to teach and you have a new approach to teacher training. This is effectively the model used by Teach For America where top-ranked university graduates are given teaching skills in a summer boot camp rather than a 4-year school of education. It is a model that could be expanded dramatically to the hundreds of university campuses that train US teachers. Take a mathematics, physics, English, or history major and teach them how to teach rather than take an education major and teach them content.

HIGH EXPECTATIONS

The greatest need in the US K-12 system is an internationally benchmarked expectation standard in the basic areas of language arts (English), mathematics, and science. Recently, 46 states have signed up for something called the "state-driven, internationally benchmarked, common core standard." This is not a national standard, but a state-driven program initiated by a CEO/Governor driven organization called ACHIEVE where each state signs up for a core standard (can be modified slightly by each state to meet local demands) such that the expectations for a child in any of the states in any grade level in any English, math, or science class are the same. Aligning the core standard to international standards automatically allows comparison to the best in the world and simplifies the task of the 15,000 US school districts in setting their own standards. The implementation of these core standards is starting now, with teachers being given training across the participating states. This is perhaps the biggest bottoms up change to the K-12 classroom in the last 100 years. If successful, it will be possible to compare any child, any classroom, and any school in the United States to the best in the world. Perhaps, the biggest challenge to this introduction of the common core standards will not be the efforts by the teachers but the realization of the general public and politicians when the initial results come in showing the poor performance of US children. Of course, this is known now but the direct comparison from the common core results will stress the system to either make needed improvements or abandon the approach. An additional challenge is the political backlash against any "national standard" due to the perception of loss of local control. Several early adopters of common standards have already back tracked due to political pressure.

TENSION IN THE SYSTEM

The US K-12 system has long been a public monopoly without accountability. This began to change with the introduction of No Child Left Behind under the last Bush Administration, which attempted to measure the performance of children and schools, especially the minority gaps, but was mired in political in-fighting and unrealistic plans. The success of NCLB was that it started the measurement process, the first step in introducing tension in the system. Today, we have many states that are moving beyond NCLB by demanding that the system measure the performance gains made by each student and relate these gains back to the quality assessments of teachers, administrators, schools, and districts. These changes are not without opposition as the monopoly establishment is fighting the changes claiming that K-12 education is different from all other enterprises, and it is impossible to measure the quality of the work force.

This movement to measure the output of education (learning by the child) has substantially changed the education reform debate. For decades, the debate has been about input parameters (size of classroom, dollars spent per child, hours in the school day or days in the school year, and the various methods of teaching) and not about the output of the process (what does the child learn). Now with new data systems, it is possible to track the yearly progress of the student and relate this back to the teacher and school and grade them accordingly. Supplementing this new found output tension has been the movement to competition from charter schools.

In over 40 states, children and parents now have the choice to send their child to a standard public K-12 school or a public charter school where the charter school is generally free to operate without many of the restrictive oversights weighing on standard public schools such as teacher certification, tenure, and rigid curriculum. School districts such as those in Washington, DC, New Orleans, Arizona, and other states have been leading the way here. Charter schools give monopoly breaking choice to the child as well as specific education focus areas such as science, arts, language, and technical training. An important measure for the success of charter schools is that they can be graded on output and closed down if they are not succeeding. This would be an important proposition for standard public K-12 schools as well but is very difficult to implement in practice.

An interesting proposition from charter school advocates is something called the "backpack funding model." In this model, every child at age 5 is given a backpack and each year the state fills up the backpack with the annual per child funding dollars for education and the child and parent have the choice to spend those dollars at whatever school best fits the child's needs and desires. This would include normal public schools, charter schools, private schools, home schooling, and so on. In this fashion, you would foster increased choice and competition within K-12 education and the participants in the process (children and parents) could decide where to spend their dollars to get the best outcome for the child.

INTELLIGENT USE OF TECHNOLOGY IN THE CLASSROOM

There has been much recent discussion about the role of technology in the education process. With the advent of online schools, hybrid online/ brick-and-mortar schools, and increased penetration of technology into the classroom, there are myriad possibilities of what education might look like in the future. Certainly, it is possible to have a world class K-12 education system without extensive use of technology. Many countries around the world demonstrate this is possible today. The question is how to best use the current technology to supplement the learning process in areas such as individual-paced learning, access to advanced learning where local content experts are not available, how to integrate the world's body of knowledge into the classroom, and how to make abstract ideas and concepts come alive in the students mind through animated learning.

Perhaps, the best way to approach this challenge is to go back to the basic learning model between teacher and student and realize that it is great teachers that get great results and not great computers. The magic in the classroom is the teacher and not the computer. Yes, the computers and internet connections are great tools for learning but the learning is directed and orchestrated by the teacher. So the challenge for the system is the intelligent use of the tool and to let the teacher use the tool in the best fashion that meets the occasion. It does not seem that technology is an alternative to what we have described earlier – great teachers, high expectations, and tension in the system. Often, we look for simple silver bullets to solve complex problems such as K-12 education. The caution here is that technology is a tool to help solve the problem but it is not the silver bullet.

MAKE EDUCATION RELEVANT FOR THE STUDENT

The general philosophy in the United States is that a K-12 education should prepare a child to be either career ready or ready to pursue a postsecondary education without remediation. Today, the system does not really succeed at either. About 25% of the students do not graduate from K-12 and of those that do about one-half are not ready for postsecondary education. A common problem for college and university administrators is that K-12 graduation requirements are not aligned for postsecondary education course work and many incoming university students are required to take remedial training. The move to the common core standards should change one of these challenges as the tests associated with the common core will be aligned to probable success by the student in university studies.

Perhaps, the biggest challenge is what to do with the 25% of the students who drop out of K-12 and do not possess the basic skills in language, mathematics, or basic science to participate in a meaningful way in the work force. A common cause of dropping out of K-12 education is the perceived lack of relevance of the education process by the child. Making education relevant is not dumbing down the curriculum but rather making sure the course offerings are aligned with future employment opportunities. For example, auto mechanics or machinists or information technology technicians need the same basic math and science courses as students preparing for postsecondary education. But students need to see that their work environment demands are consistent with the secondary education offerings. What we used to call vocational training, now renamed continued technical education, has shown great potential to keep students interested in the education process, keep them in the classroom and keep them on track to graduate from K-12. While the university-bound students clearly see the need to finish K-12 as a prerequisite to postsecondary education, the work-bound student must see the need to complete K-12 as the prerequisite for a career ready choice. Not every student may graduate from K-12 but we can do much better demonstrating to the student the importance and necessity of a K-12 education in order to be able to pursue a career of choice.

We have been debating how to improve K-12 education in the United States for over five decades. The key issues are not unknown, and there are well-documented solutions that exist from isolated examples of excellence both within the United States and from around the world. The problem to date has been not what to do, but rather, how to generate the political will to make the necessary changes. As we move forward into the 21st century, the importance of a well-educated work force for economic growth and national defense is increasing every day. We have it within our power to make the necessary changes but as with all change it will be difficult. The political and institutional forces that oppose these changes must be overcome if we expect a different result.

3

SECURE AMERICA'S ECONOMIC FUTURE BY INVESTING IN YOUNG CHILDREN

DEBORAH J. STIPEK

The United States once boasted the most educated people in the world. No longer. The comparative high school graduation rate has plunged in recent decades from the highest among OECD countries to 21st.[1] US college completion rates are also being surpassed by one country after another, and US student performance in international comparisons is mediocre, at best.

The enormous cost of losing ground in educational attainment is well understood by policy-makers. Bush's education goal was to increase academic skills to the equivalent of half a standard deviation on the PISA test used for international comparisons by 2000. Economist Eric Hanushek estimates that if that goal had been met (it was not), the GDP would have been 4.5% higher in 2015 than it was given the absence of such gains.[2]

To maintain its economic position, the United States needs to catch up with countries that are passing it by in education. One of the best levers for improving educational outcomes is to invest in children in preschool – long before they are making decisions about whether to drop out of high school or go to college. Better academic skills at school entry are associated with higher

---

[1] http://www.oecd.org/education/skills-beyond-school/48630687.pdf.
[2] http://educationnext.org/education-and-economic-growth/.

Perspectives on Complex Global Challenges: Education, Energy, Healthcare, Security and Resilience, First Edition. Edited by Elisabeth Paté-Cornell, William B. Rouse, and Charles M. Vest.
© 2016 John Wiley & Sons, Inc. Published 2016 by John Wiley & Sons, Inc.

test scores throughout K-12, higher high school and college graduation rates, and higher rates of employment and earnings.

In one analysis of a large preschool program in Chicago, children received net benefits at age 26 totaling $83,708 per participant in 2007 dollars, compared with children who did not take part in the program.[3] When projected over a lifetime, economic benefits of the program to participants and society at large amounted to nearly $11 for every dollar spent, which corresponds to an 18% annual rate of return on the original investment. Economist Tim Bartik's analyses suggest that a national program of universal, high-quality preschool education would boost the size of the national economy by almost 2% and generate 3 million more jobs and almost $1 trillion in increased annual gross domestic product. He concludes that $1 invested in preschool education would increase the present value of earnings in the nation by almost $4.

Preschool costs are partly paid for by reducing public expenditures for other services. Children who have attended high quality preschool are less likely to be retained in a grade, participate in special education, become pregnant as a teenager, or be incarcerated. All of these effects benefit the society as well as individuals, and some studies have shown that the benefits endure long into adulthood. By age 40, adults who had been randomly assigned to the Perry preschool program were more likely to have graduated from high school and remain married and less likely to have been arrested or depend on welfare programs.

Why does early childhood education yield economic benefits that are so much stronger than later investments? Children who enter school with basic literacy and numeracy skills and emotional and behavioral self-regulation can take better advantage of the learning opportunities encountered in kindergarten and beyond. As Nobel Laureate Paul Heckman points out, "skill begets skill and learning begets more learning" (Heckman & Masterov, 2004).

Neuroscience provides another explanation for the effects of preschool. Early experience and learning opportunities create the neurological foundation for later learning. The very architecture of the brain is affected, for example, by early emotional trauma and environmental deprivation in ways that undermine learning and development throughout childhood.[4] Efforts to improve the social and intellectual environment of vulnerable children can help buffer those effects.

An educational investment in early childhood education, especially for children who live in poverty, would have the added value of reducing the

[3]http://www.nih.gov/news/health/feb2011/nichd-04.htm.
[4]http://developingchild.harvard.edu/index.php/resources/reports_and_working_papers/working_papers/wp2/.

persistent achievement gap. The performance gap between the most- and least-proficient students in the United States is among the highest of all OECD countries,[5] and the correlation between academic performance and family income is rising in the US. Children from low-income families begin school at least a year behind their middle class peers and most never catch up. Entry skills are strong predictors of later academic performance and most analyses show that the achievement gap increases over the school years. By one estimate, closing the achievement gap between black and white children at school entry would eliminate half of the gap in high school (Jencks & Phillips, 1998).

Money spent on making high-quality preschool more widely accessible would yield substantial returns to individuals and to society. What is needed is political courage to take the long-term perspective required of investing in young children. In this increasingly competitive and global economy, the economic future of the United States depends substantially on its investment in human capital, and preschool is not too soon to start.

REFERENCES

Heckman P, Masterov, D. 2004. The Productivity Argument for Investing in Young Children. Available at http://jenni.uchicago.edu/human-inequality/papers/Heckman_final_all_wp_2007-03-22c_jsb.pdf. Accessed 2016 Jan 19.

Jencks C, Phillips M, editors. *The Black–White Test Score Gap.* Washington, DC: Brookings Institution Press; 1998.

RECOMMENDED READING

Bartik T. *Investing in Kids: Early Childhood Programs and Local Economic Development.* Kalamazoo, Michigan: W. E. Upjohn Institute for Employment Research; 2011.

[5]http://www.all4ed.org/files/IntlComp_FactSheet.pdf.

4

THE FUTURE OF TEACHING IN THE UNITED STATES

LINDA DARLING-HAMMOND

US education has been in a state of "reform" since a National Commission on Educational Excellence issued its report, *A Nation at Risk,* in 1983. Although the United States was the undisputed world leader in education in the 1960s and the 1970s, that status is now held by nations such as Finland, Singapore, and South Korea, which graduate about 90% of their citizens from high school and send more than three-fourths onto technical colleges or universities.

When three-quarters of the fastest growing occupations require postsecondary education, our college participation rates have declined from 1st in the world to 17th. High school graduation rates – at only 80% – have slipped below those of many industrialized nations.

What went wrong? In the last 30 years, a blizzard of initiatives has been launched to improve schools. Under President Obama, this included a federal "Race to the Top" initiative to reward states for reforms ranging from test-based teacher evaluation to starting charter schools. President Obama has set a new goal of leading the world in the proportion of college graduates by 2020. Goal setting, however, is not magic: In 1989, then-President Bush and the nation's governors set goals to graduate all students and become first in the world in math and science by 2000. Today we are further from these goals than we were then.

Perspectives on Complex Global Challenges: Education, Energy, Healthcare, Security and Resilience, First Edition. Edited by Elisabeth Paté-Cornell, William B. Rouse, and Charles M. Vest.

This troubling performance is largely because a yawning "opportunity gap" has led to one of the widest achievement gaps in the world, one that has grown in the last 20 years. The rate of childhood poverty has climbed to more than 22%, far above the rates of poverty in other industrialized nations. Furthermore, low-income children increasingly attend racially and economically segregated, under-resourced schools. These often lack basic educational materials and are staffed by a revolving door of inexperienced and untrained teachers.

Thus, while American children who attend schools with fewer than 25% of their children in poverty rank at the top of the world in reading achievement, besting even the famously high-achieving Finns, those who attend schools that serve concentrations of children in poverty, score near the bottom of OECD nations. The same gaps exist in mathematics and science, but the United States does generally less well in these subjects, and even higher-achieving American students are not internationally competitive.

Real improvement will happen only if we change our strategy – shifting from our love affair with passing fads and small-scale innovations to a focus on building a strong and equitable system that can spread best practices across all schools.

Americans are great innovators, and educators have created thousands of exciting and successful schools, programs, and projects. But these live at the margins of the system, often in hostile policy environments, and cannot be spread to other schools that lack the knowledge, resources, and capacity to adopt them. Bottom-up innovation, while necessary, is not sufficient. We must invest in a teaching and learning system, such as those in other high-achieving nations, that can routinely produce excellent schools (Darling-Hammond, 2010).

What do high-achieving nations do? First, they fund schools centrally and equally rather than allowing the dramatic inequalities common in the United States. Most states have at least a 2-to-1 ratio between their high- and low-spending schools, with schools serving poor children generally having larger class sizes, a lower-quality curriculum, and less well-qualified teachers than those serving more affluent teachers.

Second, high-achieving nations, such as Finland and Singapore, competitively select teachers from the pool of college graduates and give them a top-quality preparation, completely free, before they enter classrooms where they are carefully mentored and well-supported. Teachers earn equitable and competitive salaries, comparable to those of engineers. There are no shortages (Schleicher, 2012).

In the United States, teachers' preparation is unsupported: Most teachers must go into debt to enter a profession where they earn about 40% less than other college graduates. In the poorest schools, many enter without having finished – or sometimes even started – their preparation, earning salaries lower than those of other teachers and spending money out of their pockets to buy books and paper for students. Because of these conditions, teacher shortages, especially in fields such as math and science, are widespread.

Finally, in high-achieving nations from Finland and Singapore to Canada and Australia, curriculum and assessments focus on the higher-order skills essential in a knowledge-based economy. Guided by state or national curriculum objectives, exams emphasize essay writing, scientific investigation, research projects, and complex problem-solving – the skills demanded by universities and employers. Teachers help develop and score the assessments, which helps them understand standards deeply and continually improve their instruction. By contrast, the United States is still wedded to factory-model multiple-choice tests that are poor measures of complex skills and reinforce low-level teaching.

There are some recent openings for real reform. Recently, most states have developed and adopted new college- and career-ready standards. Intended to be "fewer, higher, and deeper" than our current mile-wide, inch-deep standards, these could guide curriculum and assessments that are more internationally competitive, and could provide more robust resources from which teachers can plan together. Consortia of states are developing new assessments that may offer more opportunities for students to demonstrate higher-order thinking and reasoning skills, giving more productive signals for instruction. A new federal law, replacing the discredited No Child Left Behind Act, offers greater levers for equitable resources and deeper learning that could strengthen instruction, especially for students in the highest-need schools.

This is just a start. To make real progress, we must build serious teaching and learning systems in every state. That means supporting states to develop high-quality curriculum and robust 21st-century assessments, creating an infrastructure for recruiting, training, and supporting excellent teachers for every community, and equalizing resources for all schools. The following steps, already in place in high-achieving nations, would be central to such an agenda:

- *Incentives to recruit a diverse and talented teaching force*, including equitable, professionally competitive salaries, and service scholarships to underwrite training, in return for teaching in high-need fields and locations;
- *Universal high-quality teacher education*, leveraged through stronger accreditation and licensing, and featuring extensive clinical training in model schools, such as teaching hospitals in medicine;

- *Performance-based licensing* – such as a bar exam for teaching – that measures actual ability to teach and creates higher standards for entry;
- *Mentoring for all beginners* from expert teachers, coupled with a reduced teaching load and shared planning time;
- *Ongoing professional learning*, embedded in 5–10 hours of planning and collaboration time at school, plus additional professional learning time to attend seminars, visit other classrooms, and conduct action research;
- *Meaningful evaluation* that combines evidence of practice in relation to professional teaching standards with multifaceted evidence of student learning, plus contributions to the improvement of the school as a whole;
- *Career ladders* that encourage expert teachers to contribute to curriculum and assessment development, mentoring and coaching for their peers, and to become school leaders, so that their knowledge and skills lift the profession as a whole.

Our challenge is to move beyond pilot projects and fleeting innovations to scale up the professional building blocks that will create a strong teaching force in every school. With that solid foundation, we can provide our children the education they deserve and the learning system that a leading nation requires.

REFERENCES

Darling-Hammond L. *The Flat World and Education: How America's Commitment to Equity Will Determine Our Future*. NY: Teachers College Press; 2010.

Schleicher, A. (2012) *Preparing Teachers and Developing School Leaders for the 21st Century: Lessons from around the World*. OECD Publishing, http://www.oecd.org/site/eduistp2012/.

5

THE CONUNDRUM OF CONTROLLING COLLEGE COSTS

Lawrence S. Bacow

Every spring on college and university campuses throughout the country, senior administrators gather to set tuition and fees for the following year. And on almost every campus, they reach exactly the same conclusion – tuition and fees must go up, typically by a rate far in excess of the rate of inflation. Predictably, the annual announcement of tuition and fee increase prompts cries of outrage from students and parents and much criticism from politicians and commentators alike. Recently, this criticism has reached a crescendo. We hear talk of a tuition "bubble," that the business model for higher education is broken, and that college and university leaders are either incompetent in their ability to rein in costs, or insensitive to the plight of students and their families who struggle to pay the increasingly high cost of a college education.

A confession. I have participated in the annual meeting described here either as a senior administrator or trustee at four separate institutions: MIT, Tufts, Harvard, and Wheaton College. I would like to think I am neither incompetent nor insensitive, yet I must admit that I failed to keep tuition growth at any of these fine institutions to anything close to the CPI. Why do college costs keep rising inexorably? And why is it that college leaders and trustees seem powerless to do anything about the problem?

Competition and costs: Higher education is an intensely competitive business. There are over 4,000 colleges and universities in the United States

Perspectives on Complex Global Challenges: Education, Energy, Healthcare, Security and Resilience, First Edition. Edited by Elisabeth Paté-Cornell, William B. Rouse, and Charles M. Vest.

and they all compete for students, faculty, resources, and visibility. In most industries, competition rewards the least cost provider. But in higher education, it seems to have exactly the opposite effect.

We actually know how to make higher education less expensive. It is not that hard. All it requires is larger classes, less student–faculty contact, less hands-on learning, a simplified curriculum, fewer student services, simpler facilities, and less support for athletics and other co-curricular activities.

The problem is that much of the market does not embrace this vision of higher education. In 10 years as President of Tufts, not once did I hear a student (or their parent) tell me that they preferred that we cut back in any of the areas described earlier and reduce tuition. If anything, the pressure was always to do more. If we had 24 varsity teams, people wanted 25. Ten study abroad programs – why not another in Africa? Shouldn't we be doing more to help our students find jobs? Why didn't we have a major or concentration in computational biology? How come the entire campus was not wireless? And while we prided ourselves on offering small, intimate classes, God forbid that any student should be closed out of a class due to limited enrollment. (My efforts to explain that we could not simultaneously guarantee both small classes and unrestricted enrollment always fell on deaf ears.)

Competition for faculty also drives costs and not just in compensation. Students are not the only ones who like small classes and state-of-the-art classrooms and laboratories. And the desire to attract and support high-profile faculty with significant research portfolios also causes institutions to make expensive investments in research facilities and infrastructure. While the social return to the nation on this investment tends to be very high, a significant part of the cost of supporting it often falls on the institution further exacerbating the cost problem.[1]

The cost disease: Another reason why costs rise faster than inflation in higher education is that productivity growth tends to lag that for the economy as a whole. In a classic study published by Baumol and Bowen (1966), they explained that in service industries with limited opportunities to substitute capital for labor, costs tend to rise faster than inflation. They characterized this problem as "the cost disease." To illustrate the point, consider the differential impact of technological change on food production and culture since the 1800s. Productivity in farming (technically output per unit of labor) has increased dramatically with the onset of mechanized farming, improvements in seeds and fertilizer, and advances in irrigation. One farmer can now produce far more food than in 1800. By contrast, it still

---

[1] Any university administrator worth his or her salt knows that research does not pay for itself due to under recovery of indirect costs.

takes four musicians to play a string quartet. As a result, the price of food has dropped relative to live chamber music. Higher education is a lot more like chamber music than farming. And it is only with the recent introduction of new online educational technology that we have the opportunity to fundamentally rethink the production function for education. More about this later.

Distinguishing cost versus price: Another factor contributing to the rapid escalation of tuition in the nation's public colleges and universities is the widespread reduction in state appropriations following the great recession. Traditionally, public institutions have benefited from healthy state support that subsidizes tuition for all attendees. As this support has declined, college administrators have had to accelerate tuition increases to not only cover a rise in costs, but to also offset the reduction in the state subsidy. Leaders of these institutions are truly caught between a rock and a hard spot. They cannot operate at a deficit, and their ability to cut costs is limited in the short run. Their only alternative is to raise tuition, often at double digit rates.

How do we do better? Einstein was fond of saying that the greatest miracle in life was compound interest. Unfortunately, when you compound costs as opposed to income, miracles turn into disasters. Many critics believe that the current rate of growth in college tuition is unsustainable. I am inclined to agree. Anger about rising college costs threatens public support for higher education more broadly. Not even the wealthiest of institutions are immune from cuts in federal support for research or financial aid. We all have an interest in solving this problem.

Technology may bail us out a bit. Recent advances in educational technology offer the tantalizing prospect that we may be able to dramatically reduce the cost of teaching large numbers of students through massively open online courses (commonly known as MOOCs.) While a full consideration of these technologies is beyond the scope of this essay, the early returns are promising.[2] However, even if technology helps to reduce the cost disease in higher education, we need to be vigilant to ensure that any net savings are devoted to moderating tuition growth and not just invested to further gild the educational lily. Productivity gains should not be subsidizing climbing walls and fancier food in college cafeterias.

If college costs are driven in part by competition, perhaps a little less of it might be desirable. If we are serious about controlling costs, perhaps we should devote some thought to areas where collaboration between institutions might result in savings for all. In some areas, legislation might be required

---

[2]For an excellent and comprehensive discussion of the potential and challenges of harnessing online education to reduce costs in higher education see Bowen (2012).

to give college and university leaders freedom to explore such possibilities without threat of antitrust sanctions.[3]

Finally, we should not forget that a healthier economy is likely to relieve the pressure of rising costs on both students and their institutions. A stronger job market will help graduates more easily and quickly retire their student loans while at the same time make it easier for state legislatures to restore funding to our public colleges and universities. In the end, we all have an interest in seeing that a college education remains within reach of everyone, and strong economy may be the most direct path to this most desirable outcome.

REFERENCES

Baumol WJ, Bowen WG. *Performing Arts: The Economic Dilemma*. MIT Press; 1966.

Bowen W. *The Tanner Lectures*. Stanford University; 2012.

[3]For example, in 1994, federal legislation eliminated mandatory retirement for university faculty. When coupled with tenure, this law functionally gave tenured faculty lifetime employment where previously they had only been guaranteed a job until age 70. Today, to get tenured faculty to retire, many institutions must offer financial incentives, further driving up costs. One solution would be to limit tenure to a fixed term, say 35 years from the date of grant. Because of competitive labor markets, no single institution could do this unilaterally. It requires collective action. But such action is unlikely given the threat of a charge of collusion under antitrust law.

6

MILITARY EDUCATION

William J. Perry

Today, the US military is the most capable military force in the world. Indeed, it may be the most effective military force – soldier for soldier – that the world has ever seen. This has not always been the case. While the US military played a decisive role in the allies' victory in World War II, a key factor in its success was its sheer size, with more than 10 million men in its Army alone, and with an unprecedented output from its defense industry. In 1944, for example, the American aircraft industry produced more than 100,000 military aircraft. Certainly, the US military fought bravely in World War II, but it is clear that the winning strategy was based on overwhelming their opponents with superior numbers. In each of the Vietnam and Korean Wars, the US military, fighting against numerically superior enemy forces, was held to stalemates at the cost of more than 100,000 American military casualties.

After the Vietnam War, the US military was held in low esteem by the American public and had exceptionally poor morale and discipline. Many officers left the military in disgust during the 1970s. And General Edward C. Meyer, Chief of Staff of the Army, famously reported to the House Armed Services Committee during a hearing May 29, 1980 that we had a "hollow army." But there were those who stayed and they were determined to rebuild a military of which they could be proud.

They decided that the key to this rebuilding was an extensive new program in military education and training. But such a program would be costly,

Perspectives on Complex Global Challenges: Education, Energy, Healthcare, Security and Resilience,
First Edition. Edited by Elisabeth Paté-Cornell, William B. Rouse, and Charles M. Vest.
© 2016 John Wiley & Sons, Inc. Published 2016 by John Wiley & Sons, Inc.

ie cost could be justified only if military personnel stayed in the force
g enough for the Army to reap the benefits of their training. Obviously, it
ade no sense to spend large sums of money to train soldiers who were to
leave the Army after two or three years of service, which had been the case
with the draftee Army.

Congress authorized the creation of an all-volunteer force, which came
into effect in 1973. Military leaders began to develop the idea of an extensive
and effective training and education program for service members and the
authorization of the all-volunteer force created the opportunity to implement
such a program.

Education and training began when a recruit entered the force and
continued throughout his career. Every promotion would depend on the
soldier having successfully completed the education and training classes
deemed necessary for his new rank. This was true for enlisted personnel as
well as officers. In time, the US Army had the best-trained noncommissioned
officer (NCO) corps in the world. And our Army took advantage of this by
passing decision-making authority down the line.

When I was Secretary of Defense, I would often take military visitors out to
one of our bases to meet our military personnel. They were always amazed by
the capability and the initiative showed by our NCO corps. Many of them told
me that they would have to go to their officer corps to get such capability. One
minister of defense actually told me that he at first believed we were deceiving
him; that he believed that the NCOs briefing him were in fact officers wearing
the uniforms of sergeants.

Our superior training programs not only gave us better-qualified NCOs,
but also gave us a major advantage in decision making in a real combat situation. Because of the quality of our NCOs and the increased authority given
to them, they did not have to pass important decisions in combat up to their
superiors. The NCO who was at the point of contact and who had the best
information on the actual combat situation was authorized and qualified to
make the necessary decisions.

In addition to the education and training programs that are part of the
regular career advancement program, our military employs special training
for specific combat situations. Many special operations forces, for example,
are trained in the language of a country in which they will be operating so
that they can operate more effectively in that country. At Fort Irwin, California, the Army has a simulated combat course where Army brigades conduct
exercises against a comparable enemy brigade, which is permanently based
there and is designed to be good enough to win most of the time. The Air
Force has comparable force-on-force exercises at Nellis Air Base. And the
Navy's nuclear submarine program has a flawless safety record, which can be

attributed to their training program – one of the most rigorous training programs in the world. Our military wants its troops to make their mistakes in training, not in actual combat. Admiral Rickover said it best: "The more you sweat in peacetime, the less you bleed in war."

Before our troops went into Bosnia, we set up a special training camp in Germany, which was built to simulate a Bosnian environment. Each battalion, before entering Bosnia, would go through a rigorous training exercise to expose them to the kind of situations they might face in Bosnia – guerilla warfare, roadside bombs, terror attacks on bases, for example. I went to that training base just before our first troops went into Bosnia to see the training just being completed. When I asked the officer in charge of the base how the training was going, he replied: "It's going fine, except that I wish the weather was colder, as it will be in the Bosnia mountain regions when they get there." This made a deep impression on me, especially since, as he was speaking, I was standing there freezing in the wind and snow.

Of course, our military has an education as well as a training component. All of our NCOs and officers have university-level courses, which they take as their career progresses. Our three military academies conduct research and offer postgraduate courses in addition to their 4-year programs for commissioning young officers. Beyond that there are graduate institutions: each service has the equivalent of the Army War College, which offers programs comparable to a Master's degree. Every field-grade officer who is selected to become a general officer, must graduate from one of those colleges, or take an alternative program at a private university. Stanford University, for example, has eight such officers in 2013, as well as eleven recent graduates of service academies who were working on a Master's degree in engineering or science. In addition, the Naval Post-Graduate School, at Monterey, has 1,800 officers working on Master's degrees in technical fields and another 700 participating via distance learning. And the Command and General Staff School at Fort Leavenworth offers a unique program in military strategy and doctrine.

This extensive education and training program pays for itself by the increased capability acquired by our NCOs and officers. And the longer these NCOs and officers stay in the service, the greater benefit our services receive. But it is also true that these programs themselves are one of the reasons that these NCOs and officers are willing to stay in the military for a full career.

One of the important ancillary benefits of this extensive program in education and training results from our military institutions accepting officers from other nations. There are many benefits to foreign officers attending US military education institutions. The foreign officers selected for American schools are usually the most promising officers in their countries – the ones who are slated to become chiefs of staff in their own military. Their knowledge of and respect for how the US military operates facilitates joint operations with their country. And for many decades the US military has not been in any significant combat operation without allies. Today joint operations are the norm, not the exception. But, just as importantly, the camaraderie among the foreign officers and their American military counterparts can play a critically important role in future cooperation among our countries. When I was Secretary of Defense, I saw many examples of this, including one that played out in the jungles of Zaire and Rwanda with the support offered to our engineering troops and me by Gen. Paul Kagame, who had just returned to Rwanda from a course at our Command and General Staff College.

So I believe that there is no doubt that this extensive military education and training program plays a key role in the effectiveness of our military today. But it is also true that these programs are costly. So it is important to look for ways to decrease the cost of the programs if that can be done without impairing their effectiveness. For many years, our military has been experimenting with ways of using computer technology to achieve that objective.

One such program, which has recently undergone very promising trials, is called the Digital Tutor. The Navy for years has trained selected recruits in the operation and maintenance of computer networks on Navy ships. They run a 35-week program at Pensacola for that purpose and each year graduate 1,422 students from that course and send them to the fleet. About 5 years ago they decided to experiment with an alternative training program using a computer-based tutor. The idea was based on the view that the best education can be achieved by the student working one-on-one with a tutor who is an expert in the field of study. But since that is too expensive to be done on a large scale, an experiment was conducted to see whether a computer could be programmed to serve as the tutor.

The program was developed for DARPA and the Navy by a Palo Alto company named Acuitus (full disclosure: the author is on the board of Acuitus). They called the program they developed "the Digital Tutor." When the development had been completed, the Navy sent a class of recruits to take a 16-week program based on the Digital Tutor. At the completion of the course, the Navy set up a competition between three different Cohorts: the Digital Tutor graduates, the traditional-class (ITTC) graduates, and Navy expert ITs (with an average of 10 years experience, drawn from 3rd Fleet).

The three cohorts competed by solving real-world, Fleet trouble-tickets, drawn randomly from a stratified set of 20,000 tickets. Each solution for each competitor was scored on a weighted, multidimensional scale that incorporated the quality of solution, the problem difficulty, the number of uncorrected mistakes made in resolving the problem, and other parameters.

The Digital Tutor graduates' aggregate score was 267, the Fleet scored 43, and the traditionally trained students scored −7 because of their high rate of mistakes. Furthermore, the Digital Tutor graduates solved 60% of the very hard problems; the other two cohorts solved none of them.

These results were remarkable. The Digital Tutor not only was a vehicle for reducing the cost of training, it was also a vehicle for improving the quality of the training. So it appears that this approach to applying computers to training programs holds considerable promise for improving training, at least in the kind of technical programs to which it has been demonstrated to date. Early in 2013, the Veterans Administration began using this same Digital Tutor teaching program to prepare veterans for jobs with computer systems companies.

Acuitus, the company that developed the Digital Tutor, will next focus on public-school mathematics, starting with Algebra and all of the mathematics that comes before. The question is how to have the same impact in the public school students as they had with young adults: creating experts in months rather than years.

In sum, training and education is a critical factor, perhaps the most critical factor, in the effectiveness of our military today. Since the end of the Vietnam War and the beginning of the all-volunteer force, our military has consistently given high priority to education and training, and the results have been impressive. But these programs are costly and, especially in technical training, not fully effective. So our military is experimenting with new ways of delivering education and training that make effective use of modern computer technology. It is clear that in today's high-tech world, the most effective military will depend on technology, not just for its weapon systems, but also for the training of its personnel who will have to operate these technically sophisticated systems.

SECTION II

ENERGY

7

INTRODUCTION

The United States was a wood economy initially. Wooden houses, wooden ships, and wood for energy, for example, heating, resulted in denuded forests throughout the Northeast. In the 1700s, whale oil was used for lighting. The first industrial use of natural gas was in New York in 1825, although natural gas had been used as an energy source for many centuries earlier. Oil was discovered in Pennsylvania in 1859. Coal was the primary source of energy in the 19th and well into the 20th century, and remains much less expensive than oil.

The biggest use of oil has been for transportation. Interstate highways enabled cars to become the major means of personal transportation. With the tremendous growth in numbers of cars and trucks, by the 1950s oil consumption came to exceed coal. As oil imports increased throughout the 1960s and 1970s, the United States was increasingly drawn into Middle East politics. This ranged from supporting oil-producing Saudi Arabia to assuring the sea-lanes of the Persian Gulf remained open. The Persian Gulf Wars (1990–1991 and 2003–2011) epitomized the strong relationship between foreign policy and energy.

Oil no longer dominates the discussions and debates surrounding energy. This is due to emergence and, albeit often slow, maturation of alternative

Perspectives on Complex Global Challenges: Education, Energy, Healthcare, Security and Resilience,
First Edition. Edited by Elisabeth Paté-Cornell, William B. Rouse, and Charles M. Vest.
© 2016 John Wiley & Sons, Inc. Published 2016 by John Wiley & Sons, Inc.

energy sources. We briefly elaborate these trends to set the stage for the contributions by energy thought leaders that follow.[1]

ENERGY DEMAND

Projections in the US Department of Energy's "Annual Energy Outlook 2015" focus on the factors expected to shape US energy markets through 2040 (DOE, 2015). "The projections provide a basis for examination and discussion of energy market trends and serve as a starting point for analysis of potential changes in US energy policies, rules, and regulations, as well as the potential role of advanced technologies."

Key results of these forecasts include the following:

- The future path of crude oil and natural gas prices can vary substantially, depending on assumptions about the size of global and domestic resources, demand for petroleum products and natural gas (particularly in non-OECD countries), levels of production, and supplies of other fuels.
- Growth in US energy production, led by crude oil and natural gas, and only modest growth in demand, reduces US reliance on imported energy supplies. Energy imports and exports come into balance in the United States starting in 2019–2028. Natural gas is the dominant US energy export, while liquid fuels continue to be imported.
- Through 2020, strong growth in domestic crude oil production from tight formations leads to a decline in net petroleum imports and growth in net petroleum product exports. In some cases, increased crude production before 2020 results in increased exports. Slowing growth in domestic production after 2020 is offset by increased vehicle fuel economy standards that limit growth in domestic demand. The net import share of crude oil and petroleum products supplied falls from 33% of total supply in 2013 to 17% of total supply in 2040 in the some cases. The United States becomes a net exporter of petroleum and other liquids after 2020 in some cases because of greater US crude oil production.
- The United States transitions from being a modest net importer of natural gas to a net exporter by 2017. US export growth continues after 2017. However, growth in crude oil and dry natural gas production

---

[1] Many of the trends summarized are presented in quotes, although most of these passages are substantial condensations of the source materials. It was easier to put quotes around whole paragraphs rather than put quotes around individual snippets woven into these paragraphs.

varies significantly across oil and natural gas supply regions, forcing shifts in crude oil and natural gas flows between US regions, and requiring investment in or realignment of pipelines and other midstream infrastructure.

- US energy consumption grows at a modest rate over the projection period, averaging 0.3% per year from 2013 through 2040. A marginal decrease in transportation sector energy consumption contrasts with growth in most other sectors. Declines in energy consumption tend to result from the adoption of more energy-efficient technologies and existing policies that promote increased energy efficiency.
- Growth in the production of dry natural gas and natural gas plant liquids contributes to the expansion of several manufacturing industries (such as bulk chemicals and primary metals) and the increased use of gas feedstocks in place of petroleum-feedstocks.
- Rising long-term natural gas prices, the high capital costs of new coal and nuclear generation capacity, state-level policies, and cost reductions for renewable generation in a market characterized by relatively slow electricity demand growth favor increased use of renewables.
- Rising costs for electric power generation, transmission, and distribution, coupled with relatively slow growth of electricity demand, produce an 18% increase in the average retail price of electricity over the period from 2013 to 2040.
- Improved efficiency in the end-use sectors and a shift away from more carbon-intensive fuels help to stabilize US energy-related carbon dioxide emissions, which remain below the 2005 level through 2040.

Exxon Mobile Corporation provides an industry perspective. "Global energy demand will be about 30% higher in 2040 compared to 2010, as economic output more than doubles and prosperity expands across a world whose population will grow to nearly 9 billion people. Energy demand growth will slow as economies mature, efficiency gains accelerate and population growth moderates" (ExxonMobil, 2012).

They continue, "In the countries belonging to the Organization for Economic Cooperation and Development (OECD) – including countries in North America and Europe – we see energy use remaining essentially flat, even as these countries achieve economic growth and even higher living standards. In contrast, non OECD energy demand will grow by close to 60%. China's surge in energy demand will extend over the next two decades then gradually flatten as its economy and population mature. Elsewhere, billions of people will be working to advance their living standards – requiring more energy."

ExxonMobil's 2040 projections, by energy sources, are as follows:

- Energy to make electricity will remain the largest component of demand and, by 2040, will account for more than 40% of global energy consumption.
- Oil, gas, and coal will continue to be the most widely used fuels, making up about 80% of total energy consumption in 2040.
- Natural gas will grow fast enough to overtake coal, with demand increasing by 60% through 2040.
- Demand for coal will peak and begin a gradual decline, in part due to costs imposed on higher-carbon fuels. In contrast, use of renewable energies and nuclear power will grow significantly.
- Gains in efficiency through energy-saving practices and technologies, for example, hybrid vehicles and natural gas power plants – will dampen growth and enable emissions to level off by 2030.

THE ELECTRIC GRID

"The U.S. electric grid is a vast physical and human network connecting thousands of electricity generators to millions of consumers – a linked system of public and private enterprises operating within a web of government institutions: federal, regional, state, and municipal. The grid will face a number of serious challenges over the next two decades, while new technologies also present valuable opportunities for meeting these challenges. A failure to realize these opportunities or meet these challenges could result in degraded reliability, significantly increased costs, and a failure to achieve several public policy goals" (MIT, 2011).

The MIT Energy Initiative's focus on the grid resulted in "a comprehensive, objective portrait of the US electric grid and the identification and analysis of areas in which intelligent policy changes, focused research, and data development and sharing can contribute to meeting the challenges the grid is facing." They concluded that "One of the most important emerging challenges facing the grid is the need to incorporate more renewable generation in response to policy initiatives at both state and federal levels."

The MIT study also concluded that "Increased penetration of electric vehicles and other ongoing changes in electricity demand will, if measures are not taken, increase the ratio of peak to average demand and thus further reduce capacity utilization and raise rates. Changes in retail pricing policies, enabled by new metering technology, could help to mitigate this problem. Increased penetration of distributed generation will pose challenges for the design and operation of distribution systems. New regulatory approaches may be required to encourage the adoption of innovative network technologies."

MIT's main recommendations emphasized the role of Federal and State governments, addressing cybersecurity threats, dynamic pricing of power, incentives for both utilities and consumers, information sharing, and increased research and development.

NUCLEAR POWER

"The pace of change is accelerating in electric power markets and the challenge facing US nuclear power stakeholders is whether or not they can adapt fast enough to remain relevant to future electric market needs. Deregulated electric markets will require future nuclear plant designs to have greater operational flexibility to complement the operating characteristics of intermittent renewable generation sources while sustaining the level of operational excellence the industry has currently achieved" (EISPC, 2013).

EISPC observes that "Along the path to commercialization, nuclear power generation encountered many of the challenges faced today by energy storage, Smart Grid, and renewable energy technologies. These include the need for government funding for demonstration projects, policy changes to promote widespread use, and a tolerance for the uncertainties associated with R&D type investments. When the government first promoted nuclear power to industry, nuclear power did not provide a solution to a problem the electric utility industry was trying to solve. It held promise to be a disruptive technology, but its capital intensive nature and unknown operational requirements caused most utilities to adopt a 'wait and see' attitude."

They conclude that "Having proven that it can safely and efficiently operate nuclear power plants, the industry now faces the greater challenge of economic competition from a resurgent natural gas industry and increased penetration of wind energy. Extremely low operating costs have traditionally insulated nuclear power plants from being concerned with power market margins. However, increasing penetration of wind, which is eligible for production tax credits (PTCs) that enable negative market prices during some hours, and the competitive nature of natural gas-fired power plants have lowered the price of power. Nuclear power plants in deregulated markets will be challenged to maintain operational excellence and identify cost-saving efficiencies under present power market conditions."

RENEWABLE ENERGY

The American Council on Renewable Energy provided a review of the state of renewable energy as of 2013 (ACORE, 2014). "Wind power accounts for 4% of the US grid, but 27%, 26%, and 19% in Iowa, South Dakota, and Kansas, respectively. Wind turbine prices and wind energy costs have dropped sharply in recent years. However, the unpredictability of the federal PTC in Congress is still forcing the industry into a boom-bust cycle. In general, zero-fuel-cost wind energy directly displaces the output of the most expensive and least efficient power plants currently operating. Of particular note, the 11 states that produce more than 7% of their electricity from wind energy have seen their electricity prices fall by 0.37% over the last 5 years, while all other states have seen their electricity prices increase by 7.79% over that period."

"Solar energy is the second-largest source of new electricity generating capacity in the US, exceeded only by natural gas. The cost to install solar fell throughout the year, with average system prices ending the year 15% below the mark set at the end of 2012. California led the pack in each market segment and saw a doubling of installations in both the residential and utility segments. Looking to 2014, California shows no signs of slowing down, particularly in the distributed generation market. Since Congress raised the Investment Tax Credit to 30% in 2006, solar companies across all 50 states have responded by investing tens of billions of dollars, growing solar installations by 3,000%, employing nearly 143,000 American workers, all while driving down average system prices by 60%."

"Geothermal power is one of the most efficient sources of electricity, with a capacity factor of 92%, meaning it almost fully uses the transmission capacity that it reserves and thus provides stability to transmission infrastructures. Geothermal technology has the ability to provide flexible power to support wind and solar generation. This helps address the intermittent nature of wind and solar power."

"Hydropower is the largest source of renewable electricity in the United States, responsible for over half of all renewable electricity generation last year and 7% of total electricity generation."

"Ethanol represents nearly 10% of our nation's fuel supply, producing over 14 billion gallons of American-made biofuels, and is poised to do much more. Ethanol has helped lower our dependence on foreign oil by 33%. Three cellulosic ethanol plants will come on line in 2014, producing the first commercially available cellulosic biofuel from corn stover and other agriculture waste. Algae bioreactors using carbon dioxide and waste water, bioreactors using wood from trees killed by pine beetles as a fuel feedstock, processes to convert the fiber from corn kernels into cellulosic ethanol, and grain sorghum and biogas from a manure digester to produce advanced biofuel are among the latest inventions seeking to become market innovations."

Other renewable sources of energy reviewed include marine and hydrokinetic energy, biomass power, waste-to-energy, and biodiesel and advanced biofuels.

The Paris-based International Energy Agency reported that in 2013, "new renewable power capacity expanded at its fastest pace to date. Globally, renewable generation was estimated on par with that from natural gas" (IEA, 2014).

They observed that "Over the medium term, renewables face a transition period. Despite strong anticipated generation growth, new generation, capacity additions and investment in renewable power are all expected to level off through 2020. Biofuels for transport and renewable energy use for heating and cooling face slower growth and persistent policy challenges."

They concluded that "Even with growing competitiveness, policies remain vital to stimulating investment in capital-intensive renewables. Scaling up deployment to higher levels would require stable, long-term policy frameworks and market design that prices the value of renewables to energy systems and increases power system flexibility to ensure system adequacy with greater variable renewables."

ROLE OF CONSUMERS

Consumer engagement can play an important role in the success of alternative sources of energy (Rouse et al., 2011). Smart Grid technologies and standards are nearing maturity, but successful deployment requires both consumer consent and participation. Communities moving to install Smart Grid infrastructure are encountering resistance, as seen in recent years in California and Maryland. The possibility of smart sensors intruding on consumers' privacy is an example of a concern. As a result, Smart Grid proponents now recognize the importance of consumer education and communication, leading to efforts such as the Silicon Valley Smart Grid Task Force and the Smart Grid Consumer Collaborative.

There are many potential approaches to enabling and motivating consumer engagement. A successful solution must impart knowledge of opportunities created by Smart Grid technology and likely market outcomes. The solution must also create financial and social incentives to engage in energy management activities. It must provide management tools that easily and efficiently communicate consumer preferences to Smart Grid systems.

In general, before they will participate, consumers must be persuaded to accept the Smart Grid value proposition. The most salient benefit of Smart

Grid systems is the potential to reduce consumption, leading to lower energy expenditures. Smart Grid technologies provide consumption information and management tools that make conservation easier. Pilot programs have demonstrated reduced consumption from 2% to 20%, depending on technology.

In addition to financial savings and ease of consumption management, Smart Grid systems will deliver social status and personal satisfaction associated with conservation, reduced environmental impact from energy generation, and better electric service reliability and performance. There is also potential profit from distributed generation and supply that will enable consumers to sell their excess power to the grid.

There are several requirements that the consumer–grid interface must meet. The system should provide real-time information on how much energy is being used, the cost of this energy, and the implications of switching appliances on or off. The elements of the user interface should allow users to communicate their preferences efficiently and accurately. Simple and comprehensible interfaces will make users feel more comfortable using the system. Data on users and outgoing information from users should be protected by the system. Finally, the system should be secured against the many different types of interruptions that can occur during communication or control of household appliances.

There are many positive trends in the development of the Smart Grid. However, a few conundrums remain, which are as follows:

- *Who Pays*: Utility companies, 3rd party companies, government, customers: Who pays for required Smart Grid investments?
- *Why Change*: Inertia to change in the utility industry versus the need for consistent regulation and innovation in energy offerings. Distributed generation forces utilities to decentralize yet extend transmission infrastructure to energy source.
- *Who Wins*: A wide variety of Smart Grid-related devices, software, technologies and companies, dynamic market versus interoperability and standardization. Competing and overlapping standard protocols must be standardized. Who will survive? Why?
- *Why Play*: Different types of consumers, for example, price-sensitive, energy-conservative, how to get them all involved? Consumers are reluctant to pay upfront costs for uncertain benefit while efficiency and conservation are antibusiness model for utilities.
- *When Safe*: Grid optimization allows more points of entry for breaches of security. Smart Grid devices and communication infrastructure

bear the risk of being attacked versus infrastructure complexity and investment.

- *Who Knows*: Dynamic pricing and optimal energy saving strategies versus energy usage data invading privacy and creating a backlash against Smart Grid implementation.
- *Where Stored*: Peak demand reduction, renewable resource versus energy storage feasibility, stability and cost.

These types of conundrums are relevant to the evolution of our energy system in terms of alternative sources of energy, how energy is delivered, and how energy is managed.

OVERVIEW OF CONTRIBUTIONS

The availability and costs of energy have long been a concern in the United States. The increasing global demand for energy derived from fossil fuels and the implications for climate change have also become concerns. The result has been a heated debate about sources and safety of alternative energy sources.

Richard Schmalensee of MIT addresses an important segment of energy use in "The Future of the US Electric Grid." He argues that two features distinguish the US electric power system: the great number and diversity of entities involved and the lack of a comprehensive national electricity policy. More specifically, about 3,200 organizations provide electricity to retail customers, with just over 200 investor-owned utilities accounting for about 66% of sales.

He reports that there is significant replacement of coal-fired base load generating capacity by gas-fired combined cycle facilities. Changes in power system design and operation and in federal transmission citing policy will be necessary to efficiently accommodate increased generation from large-scale variable energy resources, such as solar and wind. Thus, use of renewables is not just a technology issue.

New demands on the grid will include after-work charging of electric vehicles and plug-in hybrids. This could add significantly to system peak loads and requirements for new capacity, although real time, dynamic pricing could mitigate this need. Overcoming resistance to dynamic pricing will almost certainly require giving customers the ability to automate their response to price changes with a simple interface.

He notes that increased grid connectivity, via the so-called Smart Grid, will create vulnerabilities to various forms of accident and malfeasance that were not present in yesterday's power system. He concludes, therefore, that it is important that a single federal agency be given responsibility for cybersecurity throughout the US grid, along with the necessary regulatory authority.

John M. Deutch of MIT discusses "The Revolution in Natural Gas." He reports that there has been a completely unexpected explosion in reserves and production of oil and gas from shale and tight sand. Natural gas prices have declined by half, employment has increased by many tens of thousands, and there is now the expectation that North America will become an exporter, not an importer of energy.

The gap between the prices of oil and gas, with oil roughly four times more expensive than natural gas, provides a tremendous economic incentive to develop new technologies that will permit natural gas to displace higher cost liquid fuels in commercial applications. However, even where there are strong regulatory structures in place, there are important deficiencies in the scope and administration of regulations addressing the main environmental concerns of water quality, air quality, community and regional impacts, and induced seismicity. Progress would be surer and faster if industry adopted a more forward leaning policy based on measurement, reporting, and continuous improvement in its environmental management.

He indicates that greater use of natural gas will slow but not avoid the greenhouse gas emissions that contribute to adverse climate change. Finally, he observes that the geopolitical implications of the natural gas revolution greatly favor US foreign policy interests. However, sharply lower oil and natural gas prices mean lower revenue for producing countries such as Russia, Venezuela, and Saudi Arabia that has the possibility of destabilizing the economies and governments of those countries.

Richard A. Meserve of the Carnegie Institution for Science addresses "The Future of Nuclear Power in the United States." He projects a decline in nuclear power generation due to increasing reliance on natural gas, enormous capital costs, fuel disposal, and decline of public acceptance since Fukushima. Yet, he argues, nuclear power is the only significant commercially available low-carbon energy resource that can provide base load electricity.

Comparing nuclear power to natural gas, he notes that natural gas emits less carbon dioxide per unit energy than coal, but it is not an adequate long-term answer to the challenge of climate change. Furthermore, nuclear power provides insurance against gas price increases and serves to reduce volatility in energy prices, enabling more reliable planning.

If the United States turns away from nuclear power, he argues, we cannot expect to be a major player in defining the guidelines for its export and use by others. Thus, a modest near-term goal is simply to keep our civilian nuclear capability available to meet future needs. It will also be important to maintain our intellectual capital in nuclear power to provide a foundation for an eventual domestic nuclear resurgence.

Richard H. Truly, former director of the National Renewable Energy Laboratory, and Michal C. Moore of the University of Calgary, discuss "Renewable Energy: Balancing Risk and Reward." Renewable energy is derived from wind, tidal or river flows, solar radiation, gases captured from waste decomposition, or bio-based fuels. They indicate that renewable is not synonymous with free or costless.

An additional complication is that some are notably intermittent, particularly solar and wind power. Solar photovoltaics have a structural hurdle in that the sun only shines part of the day. A similar hurdle exists for biomass (on a different time scale), compounded by the technological difficulties in extracting and processing plant starch and sugar necessary for end-use.

Venture capital still awaits well-managed and strategic companies, which are determined to relentlessly introduce low-cost innovations and can compete with little or no subsidy. However, the biggest challenge to renewable resources may be that of abundant and low-priced gas available in virtually every corner of the world, which has the potential to temporarily nudge out renewables as the lower carbon alternative to coal

REFERENCES

ACORE. *The Outlook for Renewable Energy in America*. Washington, DC: American Council on Renewable Energy; 2014.

DOE. *Annual Energy Outlook 2015 with Projections to 2040*. Washington, DC: US Department of Energy; 2015.

EISPC. *Assessment of the Nuclear Power Industry: Final Report*. Burlington, MA: Navigant Consulting; 2013.

ExxonMobil. *The Outlook for Energy: The View to 2040*. Irving, TX: Exxon Mobil Corporation; 2012.

IEA. *Renewable Energy Mid-Term Market Report 2014: Market Analysis and Forecasts to 2020*. Paris: International Energy Agency; 2014.

MIT. *The Future of the Electric Grid*. Cambridge, MA: Massachusetts Institute of Technology; 2011.

Rouse WB, Clear T, Mao Y, Moreland D, Pradhan A, Yim J-B, Yu Z. *Enabling and Motivating Consumers to Manage Their Energy Consumption*. Atlanta, GA: Tennenbaum Institute, Georgia Institute of Technology; 2011.

8

THE FUTURE OF THE US ELECTRIC GRID

RICHARD SCHMALENSEE[1]

Since Thomas Edison's Pearl Street generating station opened in 1882, the US electric power system has grown into a vast physical and human network providing service to hundreds of millions of consumers. It has incorporated several generations of new technology and has improved its performance accordingly.

Every expectation, though, is that change will be – and will need to be – more rapid in the next few decades than in the recent past. There are new challenges and new technologies available to meet them. *Public policy will play a major role in determining whether the US electric power system continues to perform well.*

SYSTEM ORGANIZATION

Internationally, the US electric power system is distinguished by two features: the great number and diversity of entities involved and the lack of a comprehensive national electricity policy, leading to major differences in regional

[1] This brief essay draws heavily on data and results presented in the 2011 MIT Energy Initiative study *The Future of the Electric Grid*, of which I was a co-chair, but only I am responsible for the opinions expressed in this essay and any errors it may contain.

Perspectives on Complex Global Challenges: Education, Energy, Healthcare, Security and Resilience, First Edition. Edited by Elisabeth Paté-Cornell, William B. Rouse, and Charles M. Vest.

structures. These are not unrelated: institutional diversity has made it difficult to develop political support for national policies.

Investor-owned firms account for about 84% of US electric generation. The rest is produced by an array of federal agencies, cooperatives, and systems owned by state and local governments. Several hundred entities own parts of the transmission system, with government enterprises and cooperatives accounting for about 27%. The United States currently has 68 balancing authorities (the entities responsible for balancing generation and load in real time within a specified geographic region). This reflects the sector's complex history: there are eight in Arizona, but only one in New York. Finally, about 3,200 organizations provide electricity to retail customers, with just over 200 investor-owned utilities accounting for about 66% of sales.

Traditionally, electric power was mainly provided by vertically integrated government-owned or government-regulated monopolies. Beginning in the 1980s, however, it came to be recognized that competition was possible in generation, if not in transmission or distribution. Ownership of generation could be separated from the rest of the system, and an independent entity could operate the transmission system and administer a wholesale electricity market. The provision of distribution services would remain a regulated monopoly, but there might be competition in retail sales of electricity. This market-oriented model has been adopted throughout the European Union, in much of Latin American, and in many other nations – even including the Russian Federation.

The US Federal Energy Regulatory Agency (FERC) has taken steps to enable movement to this new model. But it has lacked either the authority or the will (depending on whom you ask) to implement an effective, national pro-market policy. Today, organized wholesale electricity markets cover two-thirds of the US population and meet about two-thirds of demand. In the southeast, the traditional vertically integrated utility model remains dominant with a significant federal role in some states, while in the west, particularly the Pacific Northwest, federal, municipally owned, and cooperative enterprises play important roles in the industry.

BULK POWER

Despite occasional assertions that we have a "third-world grid," *the US bulk power system – generation and high-voltage transmission – today operates reliably and efficiently.*

While demand is expected to grow slowly over the next few decades, the shale gas revolution and tightening environmental regulation are likely to lead to significant replacement of coal-fired base load-generating capacity by gas-fired combined-cycle facilities. In the absence of major changes in national policy, nuclear power seems likely to play a declining role in the electric power industry, with little or no new capacity coming on line and significant retirements slated to occur shortly after 2040.

Properly utilized, new technologies can reduce the likelihood of widespread bulk power system failures. In particular, Phasor Measurement Units (PMUs) can provide frequent, time-stamped information on the state of the system at multiple locations that can be used to detect system stress and facilitate effective response to it. The 2009 Stimulus Bill funded considerable investment in PMUs, but PMU data need to be shared over wide geographic areas and used to develop new systems that can translate the vast amount of PMU data into useful information and to integrate these and other new devices effectively into system operations. *Recent experience suggests that widespread sharing of PMU data and development of systems to use those data effectively may not happen without government intervention.*

Changes in power system design and operation and in federal transmission siting policy will be necessary to efficiently accommodate increased gener-ation from large-scale variable energy resources (VERs), particularly wind and solar generators. Increased reliance on these and other renewable energy sources has been and is likely to continue to be a goal of both state and federal policies. The greater the importance of VERs in any system, the greater the changes in system configuration and operation, with associated costs, that are required to adjust to the uncertainty and variability of VER supply. Sufficient reserve generating capacity and other sources of flexibility must be provided to enable the system to respond to output fluctuations on a wide range of time scales. (Grid-level storage seems unlikely to become an economic solution to this problem for at least several decades.)

The shale gas revolution has made it cheaper to provide flexibility using gas turbines, but existing wholesale market designs do not provide incentives for its provision. Moreover, as the electric power system becomes increas-ingly reliant on natural gas, growing tensions between the electricity and gas markets in the short run (e.g., synchronizing scheduling) and in the long run (e.g., financing pipeline expansion) will need to be resolved by participants in both markets.

Consolidation of some of the balancing areas mentioned here could reduce the costs of VER integration by enabling wider geographic averaging of VER outputs and could perhaps bring other efficiencies. A variety of institutional obstacles stand in the way of such consolidation in some areas.

Because the best wind and solar resources are often located far from major load centers, increasing the use of VER generation is likely to require construction of more transmission lines that cross state boundaries and/or the 30% of US land that is controlled by federal agencies. In contrast to interstate natural gas pipelines, for which the FERC has had siting authority since 1938, every affected state and federal agency must approve a new transmission line, and states' interests often conflict. *Federal legislation giving FERC "backstop" siting authority (so that it could approve a proposed transmission line if at least one involved state had done so but another had blocked it) would lower the cost of using more wind and solar power.*

DISTRIBUTION

Increased penetration of distributed generation and of electric vehicles pose problems for distributions systems. New technologies and regulatory changes can solve them and enhance the overall efficiency of the system.

The costs of distribution and transmission networks are largely independent of usage, at least in the short run, but they are nonetheless generally recovered through volumetric charges – that is, charges per kilowatt-hour of use – particularly for residential customers. This system eliminates a distribution utility's incentive to accommodate distributed generation or to support energy-efficiency initiatives because both would reduce its sales and profits. And consumers who reduce their usage through increased efficiency or installation of distributed generators shift their share of fixed network costs onto other consumers, who are generally less wealthy. To correct these perverse incentives, *fixed network costs should be recovered primarily through fixed charges that do not vary with current consumption.* To the extent possible, these charges should reflect each customer's contribution to the need for local distribution capacity.

While charging electric vehicles and plug-in hybrids is not likely to strain any regional bulk power system, distribution lines and transformers could become overloaded in some neighborhoods. In addition, after-work charging could add significantly to system peak loads in some areas and thus require new generating capacity. Barring on-peak charging would eliminate the need for additional capacity but would likely be unacceptable to consumers. A more efficient – and probably more acceptable – solution would be to provide incentives for charging at off-peak times when the system has excess capacity.

More generally, facing customers with electricity prices that reflect actual system marginal cost in real time, so-called dynamic pricing, can significantly reduce the costs of electric generation and the need for new capacity. Most US customers now see prices that change at most with the season, even though the marginal cost of supplying electricity may vary twofold during a typical day and increase by a factor of 10 or more when generation or transmission capacity constraints are binding. Even many of the largest commercial and industrial users do not face dynamic pricing. Recent years have seen substantial investment in "smarter" consumer meters, so-called advanced metering infrastructure (AMI), which can receive frequent signals from the utility and measure usage over short time intervals. AMI makes dynamic pricing possible, and this is a substantial part of the opportunity for the electric power system as a whole to become "smart."

Of course, electricity in the United States accounts for only a few percent of average household budgets, and most residential consumers have no interest in watching and responding to real-time electricity prices. Overcoming resistance to dynamic pricing will almost certainly require giving customers the ability to automate their response to price changes with a simple interface. *Properly implemented, dynamic pricing will enhance system efficiency and reduce the need for generating capacity; so utilities that have already installed AMI systems should begin a transition toward dynamic pricing for all customers and publicly share data from their experiences.*

In addition to AMI, several other relatively new distribution-level technologies are being deployed. These include distributed storage, "distribution automation" technologies that can sense and respond automatically to changes in the state of the system, in some cases providing automatic "self-healing" capability when faults occur, along with associated software systems and controls that can more efficiently handle consumer complaint calls, locate outages, and allocate repair crews. While these technologies are generally expensive, they have considerable promise, and their costs will come down.

Unlike conventional distribution system investments, however, investments in these new technologies may involve greater risk to the utility. Because both regulators and regulated monopolies are often punished politically for perceived failures but not rewarded for successes, *traditional US regulatory systems tend to encourage excessively conservative investment behavior, and the United States accordingly does not lead in the use of new distribution-level technologies.* This is likely to become increasingly expensive for systems that could effectively use new technologies to manage the challenges discussed here. But there is no simple solution to the problem of excessive risk-aversion.

Standardized measures of distribution system performance are not reported for most electric utilities in the United States, and even where performance is measured, it is rarely linked directly to utility profits. Consumers would very likely benefit and the electric power system's evolution would be smoother if regulators moved to performance-based regulation, using standardized performance measures.

CYBERSECURITY

The introduction of new technologies for system control and customer metering is bringing a vast increase in data communications within the grid. Though it is essential to make the grid more reliable and more efficient, *increased grid connectivity will create vulnerabilities to various forms of accident and malfeasance that were not present in yesterday's power system.*

In the United States, the cybersecurity of the bulk power system is addressed in standards developed by the North American Electric Reliability Corporation, or NERC. NERC's jurisdiction extends only to the bulk power system, however. For investor-owned utilities, concern with cybersecurity at the distribution system level lies with state public utility commissions, and cooperative and municipal utilities are not subject to any regulatory authority on this issue. Given the increasingly interconnected nature of grid operations, from generation to distribution, such a fragmented approach is surely inefficient and potentially dangerous. *It is important that a single federal agency be given responsibility for cybersecurity throughout the US grid, along with the necessary regulatory authority.*

Even with maximum effort under improved standards and regulations, it is almost certainly not possible to provide complete protection from cyberattack to a system as complex as tomorrow's electric grid. Thus, utilities and independent system operators must look beyond prevention and become prepared to promptly and effectively recover from attacks when they occur.

Consumer anxiety over the implications for privacy of the installation of smart meters is real, and failure to deal with it effectively could put at risk many of the advantages that could be gained from AMI. Privacy is not a simple issue, however, and there are no generally accepted right answers. Nonetheless, data cross state lines, so nationwide coordination is needed among all involved parties to establish consistent privacy policies.

CONCLUDING OBSERVATIONS

Between now and 2030, the US electric grid will confront new challenges and inevitably undergo major changes. Continued adaptation by all system participants, along with some important changes in public policy, will be necessary to smooth the path forward.

Additional research in a few key areas is likely to have substantial payoffs. These include the development of computational tools for monitoring and controlling the bulk power system, methods for wide-area transmission planning under uncertainty, processes for response to and recovery from cyberattacks, and development of ways to make customers comfortable with dynamic pricing. The industry should be able to support the modest efforts required if regulators will permit modest increases in utility R&D budgets, particularly for cooperatively funded projects.

Finally, it is useful to recognize that from the point of view of most residential consumers, the US grid in 2030 is not likely to differ much from the grid today. They will very likely be able to get electricity on demand almost always. They may face prices that vary over time, but their response to such prices will be largely automated. *Behind the scenes, the grid of 2030 will be much more automated and data-intensive than it is today, but turning on a lamp will produce exactly the same result as it does today.*

9

THE REVOLUTION IN NATURAL GAS

John Deutch

Five years ago, it seemed that the United States would remain hopelessly dependent on imported oil, and experts were projecting a similar trend for natural gas. Expensive imports of liquefied natural gas (LNG) were expected to be required indefinitely to replace the diminishing production of domestic conventional natural gas. Instead, there has been a completely unexpected explosion in reserves and production of oil and gas from unconventional resources – shale and tight sand – now approaching 30% of domestic natural gas consumption. Rotary steerable drilling and hydraulic fracturing, technical advances largely due to private industry, have made this increase possible. Drilling pads operating in many states now support several wells drilled to significant depth with lateral pipe extending up to 5 km. Millions of gallons of water, mixed with sand and some chemical, are injected at high pressure through precise perforations of the wellbore, creating fractures in the rock that stimulate production of the trapped fluid.

The benefits to the United States are enormous. Natural gas prices have declined by half, lowering the cost of home heating for consumers. Employment has increased by many tens of thousands, and the now the expectation is that North America will become an exporter, not an importer of natural gas (and oil) in the coming decades. Natural gas is rapidly replacing coal in the electricity sector, which brings many environmental advantages, especially significantly lower carbon dioxide emissions per kilowatt-hour of

Perspectives on Complex Global Challenges: Education, Energy, Healthcare, Security and Resilience,
First Edition. Edited by Elisabeth Paté-Cornell, William B. Rouse, and Charles M. Vest.
© 2016 John Wiley & Sons, Inc. Published 2016 by John Wiley & Sons, Inc.

electricity. Natural gas is an important feedstock for the chemical industry so its availability and low cost are a boon to this sector.

The tremendous growth in unconventional oil and gas production has begun in North America. Drilling rigs and drilling activity is much greater in the United States and Canada than anywhere else. In the United States, unlike any other country, underground oil and gas resources belong to the private landowner rather than the state, which gives exploration and production tremendous private financial incentive. Hydraulic fracturing and field practice know-how resides with investor-owned oil companies that are aggressively pursuing unconventional oil and gas opportunities, while national-owned oil companies remain focused on their abundant conventional domestic reserves. However, shale and tight sand resources are widely spread around the world: in Latin America, North Africa, China, Russia, and Europe. Over time, there is likely to be a massive increase unconventional oil and gas production that will increase the diversity and supply of these resources.

There are two key economic uncertainties. Presently, there are three natural gas markets in the world, where the price of natural gas is wildly different: $4 per thousand cubic feet (MCF) in North America, $10 per MCF in Europe, and $15 per MCF in Asia. The abundance of natural gas may well lead to greater trade through LNG tankers and progressively longer pipelines that could lead to a global market and price of natural gas such as now exists for oil. A global market price for natural gas would dramatically change the nature of energy markets, especially in the large emerging economies in Asia.

There presently exists the largest gap in history between the price of oil and the price of gas on an energy equivalent basis in North America: $90 per barrel oil is roughly four times more expensive than natural gas at $4 per MCF. Thus, there is a tremendous economic incentive to develop new technologies that will permit natural gas to displace higher cost liquid fuels in commercial applications. The transportation sector presents the greatest opportunity, as compressed natural gas would compete favorably against motor gasoline fueled light-duty vehicles. Many firms are exploring "gas to liquids" technology that converts methane into liquid fuels. It is most unlikely that the present discrepancy between the price of natural gas and oil can be sustained for many decades in the face of technical change.

The geopolitical implications of the natural gas revolution greatly favor US foreign policy interests. The increase in low-cost unconventional oil and gas resources shifts market power from supplier to consumer countries. North America with its expanded production levels and potential as a net exporter will have much greater influence in world oil markets. Many countries that possess the largest proven reserves of conventional natural gas, such as Russia and Iran, are experiencing a dramatically adverse wealth effect. Sharply lower oil and natural gas prices mean lower revenue for producing countries such as Russia, Venezuela, and Saudi Arabia that has the possibility of destabilizing the economies and governments of those countries. Friends, such as Canada and Australia, that have undertaken high-cost oil and gas production projects, may also suffer. In sum, natural gas markets around the world will experience rapid and large changes, which will require companies to be agile and states to adapt diplomacy and trade policy rapidly.

The greatest threat to expansion of unconventional oil and gas production in North America and many other regions of the world is public opposition to real and perceived adverse environmental impacts of this activity. Presently, opposition and support is evenly split in the ten or so states that have significant drilling underway. All stakeholders understand the importance of strong regulation to protect public health, safety, and environmental quality. But even in the United States and Canada where the strongest regulatory structures are in place, there are important deficiencies in the scope and administration of regulations. Comprehensive regulations must address the main environmental concerns: (i) water quality, (ii) air quality, (iii) community and regional impacts, and (iv) induced seismicity. In the United States, serious tensions exist between the federal Environmental Protection Agency and the state regulatory agencies concerning responsibility for regulations and for inspection and enforcement. Since it is possible that hundreds of thousands of hydraulically fractured wells will be drilled in the next couple of decades, it is important that regulations be crafted and administered in a way that gains the public's confidence. Furthermore, other countries are sure to look to the United States to lead in environmental regulation. Hydraulic fracturing technology is advancing rapidly. Companies and regulators are accumulating valuable field experiences. So it is reasonable to believe that industry performance on key environmental indicators will improve over time. Progress would be surer and faster if industry adopted a more forward leaning policy based on measurement, reporting, and continuous improvement in its environmental management.

The prospects for much greater production of natural gas (and oil) from unconventional sources has potential to bringing significant economic and security benefits to much of the world. Realizing these benefits requires disciplined attention to managing undesirable environmental impacts. But other challenges remain: in particular, greater use of natural gas will slow but not avoid the greenhouse gas emissions that contribute to adverse climate change.

10

THE FUTURE OF NUCLEAR POWER IN THE UNITED STATES

RICHARD A. MESERVE

Approximately 20% of US electrical energy is provided by the country's 99 operating nuclear power plants. Many of these plants will come to the end of their lives by the 2030s absent license extentions. In the meantime, few plants are being built. Four units of advanced design are now under construction in South Carolina and Georgia, but it is not known when or even whether others will follow. Given the expected growth in electrical demand and the absence of new construction, the contribution of nuclear power to our energy supply will diminish over time.

There are many reasons for this set of circumstances. Perhaps, the most important factor is the expanding role for natural gas. The price for natural gas is at record low levels and estimates of domestic gas reserves are far larger than those of just a few years ago as a result of the advent of technology to extract gas from tight shales. Because the capital cost of a gas-generating unit is far cheaper than a nuclear plant of equivalent power, the low price of fuel makes gas-powered electricity an attractive economic alternative to a new nuclear plant. Indeed, natural gas is displacing nearly all other generation sources in the absence of special policies (e.g., mandates for renewables).

Another factor that disfavors nuclear power is that the capital costs for a nuclear unit can challenge the financial capability of even the largest generating companies. The risk of cost overruns or delays can impose a crushing

Perspectives on Complex Global Challenges: Education, Energy, Healthcare, Security and Resilience,
First Edition. Edited by Elisabeth Paté-Cornell, William B. Rouse, and Charles M. Vest.
© 2016 John Wiley & Sons, Inc. Published 2016 by John Wiley & Sons, Inc.

financial burden that strongly discourages a nuclear investment in the absence of a cost-recovery mechanism through ratemaking. A new nuclear plant, with a price tag of $6 billion, is a bet-your-company transaction. Because there are other alternatives, few boards of directors have any reason to take the risk of nuclear construction in states providing competitive markets for electricity.

The failure of the United States to resolve the means to handle used fuel has also undermined public confidence. Although the National Academies have concluded that deep geologic disposal can isolate used fuel from the environment for necessary long periods of time, the long-standing controversy over disposal has caused many to doubt that we have the political capacity to make the necessary decisions to handle nuclear wastes.

Public acceptance of nuclear power is another concern. Although most polls show that a majority of Americans support the use of nuclear energy, public acceptance has understandably declined in the aftermath of the Fukushima accident. In fact, we should learn lessons from Fukushima that will make a major nuclear accident less likely. The metrics of safety performance – such as unplanned shutdowns of reactors ("scrams"), unavailability of safety equipment, radiation releases, and exposure of workers – have improved markedly over the last few decades. At the same time, advanced analytical techniques and careful evaluation of operational experience have provided deep knowledge of nuclear risks. The new reactor designs reflect this knowledge, with the result that they should be even safer than existing plants. Nonetheless, there is a deep and abiding fear of radiation that will always constrain enthusiasm for nuclear power among a segment of the populace.

There are reasons to be concerned about the gradual eclipse of nuclear power in the United States. First, as our recent weather anomalies have reinforced, there is a need to expand reliance on carbon-free energy. Nuclear power is the only significant commercially available low-carbon energy resource that can provide base load electricity. Indeed, it represents more than 60% of our current carbon-free generation. Although natural gas emits less carbon dioxide per unit energy than coal, it is not an adequate long-term answer to the challenge of climate change. Safe nuclear power should be an integral part of the response and the diminution of reliance in nuclear power takes us in the wrong direction.

Second, nuclear power provides valuable energy diversity, particularly in a nation that is becoming increasingly dependent on natural gas. Much of the cost of nuclear power arises from the predictable amortization of the plant, which means that nuclear power promises reliable energy supply at stable prices over the 40- to 60-year plant life. Moreover, although we now benefit from low natural gas prices, no one can predict whether they will remain at the current levels. Nuclear power provides insurance against gas price increases and serves to reduce volatility in energy prices, enabling more reliable planning.

Finally, there is an international dimension that should be considered. Major new nuclear construction is underway in China, India, and Russia. Moreover, the IAEA has indicated that roughly 60 countries without present reliance on nuclear power have explored whether to acquire a plant and it estimates that perhaps 15 such countries will proceed with construction over the next decade. Although the United States may retreat from nuclear power, many other countries have the opposite energy destination. There is a national security aspect to this growth in nuclear power abroad. Although nuclear power plants are not a proliferation concern by themselves, other parts of the fuel cycle present risks. More power plants necessitate more enrichment capacity and the enrichment technology used to produce reactor fuel can also be deployed to produce weapons-usable highly enriched uranium. More reactors also means there will be spent fuel in more places that could be reprocessed to extract weapons-usable plutonium. If the United States turns away from nuclear power, we cannot expect to be a major player in defining the guidelines for its export and use by others. There thus are important national-security consequences that are associated with a US retreat from a nuclear power at a time when much of the rest of the world is turning to it.

How should we confront this dilemma? In the long term, it is inevitable that this country – indeed, the world – will develop policies to address climate change. We will likely turn to nuclear power, along with low-carbon renewables and efficiency gains, as a means to respond. The current failure of the market to price the consequences of carbon emissions will have to be corrected through a carbon tax, emissions controls, clean energy credits, or some other means. But because there is as yet no political or popular realization of the need for action along these lines, a more modest near-term goal is simply to keep our civilian nuclear capability available to meet future needs. A revived loan guarantee program, tax incentives (e.g., accelerated depreciation, tax credits), and the inclusion of nuclear power in clean energy mandates could perhaps help to overcome the financial barriers to new domestic construction. And encouragement of US participation in international markets through flexible international agreements, sensible export policies, and export financing could help to keep US capabilities alive through sales abroad. The encouragement of small modular reactors may also provide an important market niche for US vendors.

At the same time, it will be important to maintain our intellectual capital in nuclear power to provide a foundation for an eventual domestic nuclear resurgence. Research to improve light water technology, to pursue advanced technologies, and to advance safety should be an element of our national strategy. The maintenance of our educational capacities will also be necessary, and could provide a partial means to maintain US influence during the hiatus in new construction.

The bottom line is that we should prepare now for the day when the resumption of nuclear construction will become a national and international priority. We should preserve US capacity in a technology where we once were the pacesetters for the world.

11

RENEWABLE ENERGY: BALANCING RISK AND REWARD

RICHARD H. TRULY AND MICHAL C. MOORE

Energy is vital for all societies, advanced or developing. For centuries, this has meant utilizing fossil fuels and more recently nuclear power. Worldwide, nations devote scarce resources to exploit so-called *renewable energy*, and companies strive to simultaneously reduce costs and apply innovative technologies to compete in this growing energy sector.

Renewable energy is derived from wind, tidal or river flows, solar radiation, gases captured from waste decomposition, or bio-based fuels. These sources tend to be naturally recharged or replenished in a seasonal or sometimes daily flux. Depending on geographic location, they work independently or cooperatively with more traditional energy resources, and in the process, create substantial public support.

A Yale – George Mason 2012 survey indicated that "majorities of Americans say that global warming and clean energy should be among the nation's priorities, want more action by elected officials, corporations and citizens themselves, and support a variety of climate change and energy policies." These majorities held true regardless of respondents' political stripe.

Adding to this public support, policy-makers and regulators have consistently shown a preference for including renewable energy resources in the electric power mix. This has held true even when technologies are less-than-competitive, because renewables enhance grid reliability through supply

Perspectives on Complex Global Challenges: Education, Energy, Healthcare, Security and Resilience,
First Edition. Edited by Elisabeth Paté-Cornell, William B. Rouse, and Charles M. Vest.
© 2016 John Wiley & Sons, Inc. Published 2016 by John Wiley & Sons, Inc.

diversity, utilize regional or local renewable supplies, and provide a hedge against a future that demands lower carbon generation across-the-board.

That said, even though these growing technologies are derived directly from nature, *renewable* is not synonymous with *free* or *costless*; an additional complication is that some are notably intermittent, particularly solar and wind power. Even biogas from plant stock may vary widely between seasons. Progress in providing effective energy storage technologies remains a major hurdle. As utility companies have gained valuable experience in using renewables, they have quickly learned that careful attention to systems integration, technical support, and reliable dispatch to consumers is vital to success.

While renewable energy sources vary widely, they commonly share a more benign environmental footprint and operating cost envelope than fossil fuel technologies such as gas- or coal-fired turbines. Initial capital costs can be steep, however. This has led to a variety of programs designed to make them more competitive through public support such as preferred dispatch, mandatory portfolio standards for utilities, feed-in tariffs and production cost, or consumer demand subsidies.

Two popular and widely distributed resources – solar photovoltaics and biomass – can be used to illustrate fundamental attributes of most renewables. Together, they define the two broad camps of energy needs; photovoltaics provide electricity as an energy carrier while biomass serves as our surrogate for liquid transportation fuels.

An enormous amount of solar energy is available everywhere on Earth but we cannot capture all of it; similarly, plant mass stores much solar energy but binds it up tightly, releasing only a fraction unless we apply much external energy to get it out. This fractional potential, that is the net production of energy from technological devices, tends to make the delivered cost of energy relatively high compared with thermal energy sources. Innovation is closing this efficiency gap, yet a technology such as solar photovoltaics still has a structural hurdle in that the sun only shines part of the day. On a different time scale, a similar hurdle exists for biomass, compounded by the technological difficulties in extracting and processing plant starch and sugar necessary for end-use.

Overcoming the inherent intermittency of renewable resources points out a key to a productive future of all innovative technologies; development must be simultaneously matched by an aggressive, inventive emphasis on across-the-board reduction in demand. Luckily, the relentless pursuit of energy efficiency has few downsides since it improves our national, business, and personal productivity that in aggregate underpins our national economy, competitiveness, and security. Make no mistake, this is high stakes competition.

The solar industry has recently undergone a wrenching but expected shake-out and has seen numerous bankruptcies. Some attribute this to low-cost international competition, others to poor planning and/or economic mismanagement. Furthermore, price competition from thermal resources, especially currently abundant natural gas, can give pause to investors interested in this sector. However, venture capital still awaits well-managed and strategic companies, which are determined to relentlessly introduce low-cost innovations, which can compete with little or no subsidy.

Renewable energy resources abound, but are not always accessible or affordable. However, with aggressive and consistent policy initiatives, including strategic research and development, renewable energy technology investment and deployment has provided attractive returns for business and consumers, illustrated in several areas:

1. In a well-managed company, returns on investment can be high and consistent over time. Initial high cost of capital tends to be balanced by low annual cost of operations. Past investment encourages governments to maintain support of the installed base.

2. The contribution to environmental quality and carbon management schemes is significant and helps offset fossil carbon generation and mitigate undesirable consequences.

3. The variety of renewable technologies has tremendous room to grow, both in terms of performance and substitution for older, less competitive generation.

Naturally, tastes, behavior, political commitment, and costs change over time, sometime radically. While renewable resources contribute to these changes, their biggest challenge may be that of abundant and low-priced gas available in virtually every corner of the world, which has the potential to temporarily nudge out renewables as the lower carbon alternative to coal.

The energy landscape is evolving and presents both a formidable challenge and a unique opportunity. Fossil, nuclear, and renewables are all credible players in the mix that benefit when used as part of an overall suite of energy resources. This theme reminds us that while energy is a necessary part of our society, it depends on a myriad of other supportive programs and policies such as land use, air and water quality limits and ultimately, the volume and nature of our overall energy demand.

SECTION III

HEALTHCARE

12

INTRODUCTION

Breakthroughs in medical science along with innovations in clinical practices and related technologies offer enormous opportunities for impressive improvements in the health and well being of society. Returns on investments in these endeavors have the potential to be substantial, sustainable, and broadly beneficial. However, we will not realize the greatest returns with our current fragmented system of healthcare delivery (Reid et al., 2005; Rouse & Cortese, 2010; Rouse & Serban, 2014). Many see a need to engineer or design a system that can provide high-quality, affordable healthcare for everyone (PCAST, 2014). Engineering healthcare delivery will require that the current nondesigned enterprise be substantially transformed (Rouse, 2006).

This section proceeds as follows. We first discuss the nature of the problems associated with the current healthcare delivery system in the United States. This includes the forces driving needs for change, the types of issues that need to be addressed, and the complexity of these issues. We then address the objectives that transformation should be designed to achieve in terms of high-value healthcare, including definitions of value, how it is delivered, and the roles of information and incentives in providing value. We next consider alternative approaches to solving healthcare delivery problems, contrasting bottom-up and top-down approaches, and the use of quantitative approaches to explore a wide range of solutions. Finally, we provide an overview of thought leaders' contributions to this section.

DRIVING FORCES

Why is the transformation of healthcare delivery so important now? There are several drivers. Insurance reform will surely be followed by healthcare reform. The on-going health policy agenda is shifting the emphasis from simply covering more people under the Medicaid and Medicare programs to changing delivery practices. The range of changes included in future healthcare reform is highly uncertain. Providers and payers need to be able to consider very different hypothetical scenarios. All in all, the Affordable Care Act has created both hope and much apprehension (Goodnough, 2013).

Employers' economic burden for providing healthcare – due in part to cost shifting from Medicare and Medicaid patients – is unsustainable (Rouse, 2009, 2010a). This hidden tax leads to competitive disadvantages in the global marketplace (Meyer & Johnson, 2012). Employees' unhealthy lifestyles are increasing the incidence and cost of chronic diseases, leaving employers to absorb both increased healthcare costs and the costs of lost productivity (Burton et al., 1998).

Healthcare providers will have to adapt to new revenue models. For example, they may be paid for outcomes rather than procedures as currently attempted under some stipulations of ACA. Improved quality and lower costs will be central. Providers will have to differentiate good (profitable) and bad (not profitable) revenue, which means that they will need to understand and manage their costs at the level of each step of each process.

Responsibility for outcomes will lead to a more networked organization to enable accessing the most cost-effective capabilities needed to assure outcomes (Shortell & Casalino, 2008). The most expensive asset, the physician, will increasingly be focused on the most complex activities, leaving other less-expensive professionals to perform functions requiring less training. Contracting, partnering and managing such a network model will, therefore, be increasingly central – and more risky in the sense that outcomes will be at risk. Without acceptable outcomes, there may be little, if any, revenue.

The transformation of health delivery will involve many issues at all levels of the system. Decisions associated with these issues should be evidence-based in the sense that data and analytics should be central to decision making. The tendency to base decisions on anecdotal experiences with the old system will wane and more rigorous approaches will be adopted.

What challenges will be central to transforming the delivery system? The overarching difficulty concerns how best to organize in response to the driving forces summarized earlier (Reid et al., 2005; Rouse & Cortese, 2010; Rouse

& Serban, 2014). It is clear from the best performers in health delivery, as well as many other domains, that a superior way to approach this issue is to focus on the processes whereby health is delivered to people. At a high level, these processes include prevention and wellness, out-patient chronic disease management, and in-patient care delivery.

Thinking in terms of processes is different from thinking in terms of departments and functions or specialties. A process orientation focuses on how value is provided to those receiving health services. Value is concerned with health outcomes, service prices, and service levels. Understanding and paying attention to processes causes providers to see themselves in terms of value streams or networks that create desired outcomes for customers with acceptable prices and service levels.

Given this process orientation, central issues concern mapping and optimizing careflow processes. This includes deciding about the sequencing and timing of process steps, as well as the allocation and scheduling of capacity to them. Since not all people have the same needs, decisions must be made about stratifying patient flows by tailoring them to risk levels, as well as creating the means of reducing risks (e.g., wellness).

There are also issues surrounding scaling of the delivery system. This includes ramping up new offerings from what worked for a pilot cohort to a much larger patient population. Related decisions concern the extent to which customized plans can be delivered by standardized processes. In other words, how can customization be delivered at scale?

Changing revenue models present substantial challenges. Adapting to payment for outcomes rather than fee for service models means that providers have to scrutinize processes to determine where value is most – and least – added. When payers no longer reimburse costs, then providers have to understand costs much more deeply than they commonly do now. Many procedures will have to be streamlined, others will need to be delivered in a nontraditional setting or by alternative personnel, and many are likely to be eliminated or shifted to the patients via various forms of e-visits.

The impacts of reduced Medicare/Medicaid reimbursements will also require decisions about who to serve and how to serve them. Some providers already limit the number of Medicare/Medicaid patients they serve. They may also decide to use very streamlined processes, effectively providing low-end services for those who cannot afford high-end services. The many hospital providers who have closed their emergency rooms or physicians who have changed to a concierge practice model are clear indicators of this strategy.

Another class of decisions concerns optimizing employer-based programs. Such programs are increasingly focused on prevention and wellness, targeting employees with high risks of diabetes or cardiovascular disease, for example. Many employers are building in-house clinics to provide convenient low-cost care to employees, typically with the management and staffing outsourced.

COMPLEXITY OF DECISION MAKING

Why is making such decisions seen as rife with complexity? One source of complexity is the interactions among different levels of the system. Government incentives and inhibitions (e.g., regulations) affect enterprise strategies for allocating resources by providers, payers, and employers, as well as suppliers such as medical device and pharmaceutical companies. These resource allocation strategies influence the management of the universe of organizations involved across the system. Organizational management, in turn, affects process operations and health delivery. The goal is evidence-based decision making at all of these levels (Reid et al., 2005; Rouse & Cortese, 2010; Rouse & Serban, 2014).

Other sources of complexity for executives and managers at all levels include the many alternative policies on a wide range of issues, the reality of outcomes often being uncertain and/or delayed (e.g., the returns on prevention), and the difficulty of understanding higher-order consequences and exploring unintended consequences, particularly the implications for clinicians and patients. The fact that there are typically so many independent provider business entities that interact in multiple, and often conflicting, ways is one of the enormous sources of complexity (Rouse, 2000, 2008). This results in many types of poorly understood and poorly managed interactions among the aforementioned system levels.

Another complicating factor is the lack of consensus on objectives. Do we want the lowest cost healthcare delivery system that provides acceptable levels of effectiveness of detection, diagnosis, treatment, and recovery? This, of course, begs the question of what is *acceptable*. In contrast, the objective could be a high-value healthcare delivery system that provides quality, affordable care for everyone. This begs the question of what is meant by "value."

VALUE AND HEALTHCARE DELIVERY

There is currently much commentary on two things in healthcare – universal affordability and cost control. However, we do not think that people want the lowest cost, universally affordable healthcare system. We think the central issue should really be the creation of a healthcare system that provides affordable high-value care for everyone.

Value is often defined in terms of the benefits of the outcomes of expenditures, divided by the costs of the expenditures. The benefits of healthcare – from a patient's perspective – include the quality of health outcomes, the safety of the process of delivery, and the efficacy of the services associated with the delivery process (Porter & Tiesberg, 2006; Christensen, Grossman & Hwang, 2008). Beyond the individual patient's perspective, there is, however, the issue of equity of access. Thus, the variability of value delivered across society is a concern.

The benefits from the perspective of society also include the availability of healthy, productive people who contribute to society in various ways. When people are not healthy, these contributions are diminished. A recent study found that the cost of the lost productivity often far exceeds the cost of the healthcare (DeVol & Bedroussian, 2007).

For many reasons, it is likely that people will remain in the workforce longer than in the past. In the 1960s, before Medicare got started, the average age of death was 67 years; the average age at retirement was 65–66 years. By 2005, the average age of death had reached 75 years and the average age of retirement was 62 years. We are now seeing some people planning on working longer, perhaps up to 70 years or beyond. Even Social Security has delayed the eligibility age for benefits, with the retirement age for people born after 1937 increasing to 66 years if born before 1955 and up to 67 years if born later.

If people who do not retire at the usual Social Security retirement age stay on employers' insurance, then employers will have an increased interest in keeping people healthy for as long as possible and doing productive work for as long as possible until the worker retires and goes onto Medicare. Employers' insurance companies are glad to see the older employees roll into Medicare for the same reasons; when the greater costs are incurred it is the government's problem.

Indeed, this is the exact reason some have proposed that people enroll in and own an insurance product and keep it even after they retire. That way the insurance company will now have an interest in keeping you as well as possible even after you retire. A Federal Employees Healthcare Plan model would facilitate this option. This, of course, has enormous implications for Medicare (Rouse & Cortese, 2010).

Thus, there will be an increasing need to keep older people healthy because they will need to remain productive longer. More pervasive, due to the "flat world" (Friedman, 2005), will be the need to keep everyone healthy and productive. Part of the value equation, therefore, should include the productivity in the future of all the workers that do not get diabetes or have heart attacks or cancer. For this to work, our value equation will have to account for future returns from today's investments.

This broader perspective emphasizes the importance of a healthy, educated, and productive population to the well being and competitiveness of a society (Rouse, 2010b). Of course, a central debate in the United States concerns who pays for this value. Regardless of how this debate is resolved, however, it is difficult to make the case for an unhealthy, uneducated, and unproductive country.

OVERVIEW OF CONTRIBUTIONS

Denis A. Cortese and Robert K. Smoldt, former CEO and CAO of Mayo Clinic, make this argument in "How to Move Toward Value-Based Healthcare?" They assert that paying for completion of process items not only lacks correlation with patient outcomes, it also does not appear to lower healthcare costs. They argue for rewarding medical providers who actually deliver value – obtain good patient outcomes while using fewer resources than others. More specifically, they propose paying for the cost of resources used by medical centers that get the best outcomes, plus a 3–5% margin. As a consequence, the medical centers presently using more resources than needed would have strong financial incentives to become more efficient. We are seeing many elements of their proposal emerging from the responses of providers and payers to the Affordable Care Act.

Beyond getting incentives aligned with desired outcomes, we also need to get the information systems associated with healthcare delivery to provide better, more timely information with lower costs. William W. Stead of Vanderbilt University argues this in "Delivering on the Promise to Reduce the Cost of Healthcare with Electronic Health Records." He begins by outlining the problems stemming from not having the right information, in the right place, at the right time. These problems include 61% of patients fearing being given the wrong medication, pharmacists placing over 150 million phone calls annually to physicians to clarify ambiguous prescriptions, and as many as 125,000 people dying annually from nonadherence to their medications.

Stead then discusses the sources of the fragmented information systems. He indicates that the healthcare industry targets quality improvement at failure *points* in care processes as they are identified. Experts then examine the cause and re-design roles, processes, or technology to reduce the chance of failure. The change is then tested in a pilot, corrected on the experience gained through the pilot, and deployed if judged successful.

However, he argues a care process must work flawlessly *end-to-end* to improve quality sufficiently to reduce the amount of care needed and the cost of the system. He concludes that government incentives for "meaningful use" of health information technology should: (1) focus on end-to-end, rather than point, use and (2) require providers to demonstrate that they *continuously* coordinate adaptation of evidence-based practices and electronic health record decision/documentation supports to improve clinical and concurrent process measures.

Many have projected the potential payoffs as healthcare takes advantage of "big data." Elizabeth A. McGlynn of Kaiser-Permanente explores this possibility in "Big Data in Health & Healthcare: Hopes & Fears For the Future." She argues that discovery and change can be bigger, faster, cheaper, and easier in a big data world compared to a little data world. Nevertheless, spurious correlations will happen and those who use the results of big data should keep that reality in mind. Another fear, she reports, is that the data, especially in healthcare, are too "messy" and that the ways in which the data are inaccurate is not completely at random. She indicates that embracing big data analysis, as a basis for action, requires a significant change for many leaders in how they make decisions. Furthermore, numerous health-related studies have shown that insight rarely leads directly and rapidly to action. Nevertheless, she concludes that leaders in healthcare should learn what they can about the world of big data, get involved in using its methods in their setting, and seek a balance between healthy skepticism and embracing innovation.

New incentive systems and new information systems and sources will place new demands on medical education. Lloyd B. Minor and Michael M.E. Johns, of Stanford University and Emory University, respectively, consider this need in "Medical Education: One Size Doesn't Fit All." They report that the range of activities, diversity of responsibilities, scope of postgraduate training, and sheer quantity of information and new technologies in medicine has increased exponentially over the past five decades. These changes have led to complex and evolving roles for physicians that have thus far not achieved full recognition in medical school curricula. In addition, many see today's medical education as too expensive, too long, and not producing the workforce society wants in the places of need.

They argue that needed new competencies include working well as a member of teams, meeting the healthcare needs of an aging population with an increasing number of chronic diseases, addressing health disparities, and promoting quality and continuous process improvement. Advances in instructional methodologies such as online education, podcasts, new web- and app-based tools, and simulations will accelerate the development of more cohesive links between specific knowledge and skills and the different roles of physicians. They suggest that benefits that will accrue from these reforms include reductions in the length of training for some physicians, closer interactions with schools in other health professions, and reductions in the cost of medical education. They conclude that the process of achieving better alignment between educational objectives and the roles of physicians will improve the ability of physicians to meet the needs of society.

REFERENCES

Burton WN, Chen C-Y, Shultz AB, Edington DW. The economic costs associated with body mass index in a workplace. J Occup Environ Med 1998;40(9):786–792.

Christensen CM, Grossman JH, Hwang J. *The Innovator's Prescription: A Disruption Solution for Health Care*. New York: McGraw-Hill; 2008.

DeVol R, Bedroussian A. *An Unhealthy America: The Economic Burden of Chronic Disease*. Santa Monica, CA: Milken Institute; 2007.

Friedman TL. *The World Is Flat: A Brief History of the Twenty-First Century*. New York: Farrar Straus Giroux; 2005.

Goodnough A. A Louisville clinic races to adapt to the health care overhaul. New York Times; 2013 Jun 23, 1.

Meyer JA, Johnson WR. Cost shifting in healthcare: an economic analysis. Health Affairs 2012, May 23; 20–35.

PCAST. *Better Health Care and Lower Costs: Accelerating Improvement Through Systems Engineering*. Washington, DC: President's Council of Advisors on Science and Technology; 2014.

Porter ME, Tiesberg EO. *Redefining Health Care: Creating Value-Based Competition on Results*. Boston: Harvard Business Review Press; 2006.

Reid PP, Compton WD, Grossman JH, Fanjiang G. *Building a Better Delivery System: A New Engineering/Health Care Partnership*. Washington, DC: National Academies Press; 2005.

Rouse WB. Managing complexity: disease control as a complex adaptive system. Inf Knowl Syst Manage 2000;2(2):143–165.

Rouse WB, editor. *Enterprise Transformation: Understanding and Enabling Fundamental Change*. New York: Wiley; 2006.

Rouse WB. Healthcare as a complex adaptive system. Bridge 2008;38(1):17–25.

Rouse WB. Engineering perspectives on healthcare delivery: can we afford technological innovation in healthcare? J Syst Res Behav Sci 2009;26:1–10.

Rouse WB. Impacts of healthcare price controls: potential unintended consequences of firms' responses to price policies. IEEE Syst J 2010a;4(1):34–38.

Rouse WB, editor. *The Economics of Human Systems Integration: Valuation of Investments in People's Training and Education, Safety and Health, and Work Productivity*. New York: Wiley; 2010b.

Rouse WB, Cortese DA, editors. *Engineering the System of Healthcare Delivery*. Amsterdam: IOS Press; 2010.

Rouse WB, Serban N. *Understanding and Managing the Complexity of Healthcare*. Cambridge, MA: MIT Press; 2014.

Shortell SM, Casalino LP. Healthcare reform requires accountable care systems. JAMA 2008;300(1):95–97.

13

HOW TO MOVE TOWARD VALUE-BASED HEALTHCARE?

DENIS A. CORTESE AND ROBERT K. SMOLDT

Winston Churchill is said to have once quipped, "You can always count on Americans to do the right thing – after they've tried everything else." He may have been right when looking at most of the US healthcare pay for performance (P4P) efforts. To date, most such efforts have focused on financial rewards to medical providers for completing a list of process items, for example, giving a heart patient aspirin on admission to the hospital or giving a surgical patient antibiotics within 1 hour before surgery. A major problem with such an approach has been highlighted in a number of recent publications. For example, "Efforts to pay hospitals on quality didn't cut death rates, study finds" was the headline from *Kaiser Health News* on March 28, 2012. This is just one of the more recent studies reaching such a conclusion.

But there is a long history of similar conclusions. Nearly 39 years ago, the *New England Journal of Medicine* published an article by Brook and Appel, which reached the following conclusions: "It is ironic that the outcome indicators that are of vital importance to the patient ... had non-significant correlations with either implicit or explicit process judgments. ... the use of criteria lists ... may ... decrease efficiency of medical care ... without substantially affecting the health of patients."

And the United States is not the only country to come to this conclusion. Even Churchill's England has been surprised by such a finding. The

---

following, is a headline from London's *The Times* on May 12, 2003: "'Best' Hospitals Have Worst Death Rates … Patients are more likely to die [on a risk-adjusted basis] at hospitals rated as outstanding [based on process and structure measures] by government … "

Paying for completion of process items not only lacks correlation with patient outcomes, it also does not appear to lower healthcare costs. On January 18, 2012 the Congressional Budget Office (CBO) issued a report reviewing the results of ten Medicare demonstrations – nine of which defined outcomes in terms of process items. Of those nine, none reduced per patient healthcare costs and some actually increased costs.

Michael Porter, one of the world's leading authorities on management thought, has taken a hard look at how to best improve healthcare. In *Redefining Health Care*, he concludes that

> These current [P4P] efforts … carry some risks. Most … are not actually about quality results, but processes. Most 'pay for performance' is really pay for compliance … Compliance to too many process standards runs the risk of inhibiting innovation by the best providers … Overall, attempting to micro-manage hospitals and doctors by specifying processes is a difficult task that will only become a morass … The only truly effective way to address value in health care is to rewards ends, or results, rather than means, such as, process steps.

It is time to follow Porter's advice and reward medical providers who actually deliver value – obtain good patient outcomes while using fewer resources than others.

Common sense says if we would start paying for value (rather than process) we would be more likely to get it. In his book *Total Cure: The Antidote to the Health Care Crisis*, Luft outlines a way to get there. Since we are concerned about overall healthcare costs, we should start with the most expensive cohort of patients – keeping in mind that in any given year, 80% of total costs come from 20% of patients. Most of the expensive patients in the cohort are patients who have been hospitalized. Luft suggest a modification for how providers would be paid for hospitalized patients. The new payment scheme would be built on the existing Diagnosis Related Group (DRG) system in an approach he calls "Expanded DRG's" (EDRG). For each of the present Medicare DRGs, the lump sum payment would be expanded in two ways:

1. It would include all physician services as well as hospital services.
2. It would cover a longer time period than the initial hospitalization.

For example, the EDRG for a hip replacement might be for all services related to the hip from the hospital admission through the next 6 months. Former Obama Administration advisor Dr Ezekiel J. Emmanuel has explained why this approach will improve results. "The idea is to force all of the patient's care providers to work together. They have a strong incentive to eliminate unnecessary tests and treatments and to use less expensive implants, drugs and devices that don't compromise quality, and to prevent infections and other complications that could land the patient back in the hospital." It is recognized that medical centers with independent physicians will have a more difficult time adjusting to the proposed EDRG approach, but since the goal is more integrated care, the financial incentives of EDRGs will likely force a greater degree of integration between medical centers and independent physicians. Moreover, since patient compliance to treatment is critical, this payment approach will also encourage providers to find new ways of promoting patient activation and engagement in their own care.

The most critical element is how to set the lump sum payment amount. Rather than using the standard Medicare approach of complex formulas, Luft suggests " ... Pay [providers] ... [the actual cost of providers] with outcomes above the median."

In essence, this approach is *reality-based* pay for value. It should pay for the cost of resources used by medical centers getting the best outcomes plus a 3–5% margin – but it would not pay more than that. In this approach, the medical centers presently using more resources than needed would have strong financial incentives to become more efficient – to deliver better value.

It is possible to deliver better patient care at lower cost per patient than that exists on average in the United States. Indeed, many medical centers are already doing this. Why not financially encourage the other medical centers to do the same? After all, if we actually pay for value, we are more likely to receive it.

RECOMMENDED READINGS

Congressional Budget Office. Lessons from Medicare's demonstration projects on disease management, care coordination, and value-based payments; 2012 Jan 18.

Emanuel EJ. Saving by the bundle. The New York Times; 2011 Nov 16.

Jha et al. The long-term effect of premier pay for performance on patient outcomes. N Engl J Med 2012;366:1606–1615.

Luft HS. *Total Cure: The Antidote to the Healthcare Crisis*. Cambridge: Harvard University Press; 2008.

Porter ME, Teisberg EO. *Redefining Health Care: Creating Value-Based Competition on Results*. Boston: Harvard Business School Press; 2006.

Porter ME. What is value in health care? N Engl J Med 2010;363:2477–2481.

14

DELIVERING ON THE PROMISE TO REDUCE THE COST OF HEALTHCARE WITH ELECTRONIC HEALTH RECORDS

WILLIAM W. STEAD

The 2009 Health Information Technology for Economic and Clinical Health (HITECH) Act provided $50 billion to achieve electronic health records for every American by 2014. This investment was to set the stage for healthcare reform by providing an information "foundation" to support quality improvement and cost reduction.

The opportunity for improvement is clear. Sixty-one percent of patients fear being given the wrong medication. Over 150 million phone calls are placed annually by pharmacists to physicians to clarify ambiguous prescriptions. As many as 125,000 die annually from nonadherence to their medications.

These problems stem from not having the right information, in the right place, at the right time. Electronic health records are a critical piece of the solution. A 2005 study by RAND Corporation estimated *potential* savings from electronic health records of $81 billion per year.

Since 2009, hospital and physician utilization of electronic health records has more than doubled. However, the impact on healthcare cost has been

Perspectives on Complex Global Challenges: Education, Energy, Healthcare, Security and Resilience, First Edition. Edited by Elisabeth Paté-Cornell, William B. Rouse, and Charles M. Vest.

mixed as noted in the January 10, 2013 *New York Times* – "In second look, few savings from digital health records."

The gap between the potential to increase healthcare value and what is being achieved is not surprising. Causes of this value gap are part sociocultural, part technical, and part related to *how* the healthcare industry approaches quality improvement. Although the sociocultural and technical aspects will take time to change, the approach to improvement could change quickly, buying time for the others to follow.

The healthcare industry targets quality improvement at failure *points* in care processes as they are identified. Experts examine the cause and redesign roles, process, or technology to reduce the chance of failure. The change is then tested in a pilot, corrected on the experience gained through the pilot, and deployed if judged successful. The focus then shifts to the next failure point. Over time, this approach reduces failure points and improves quality, but not enough to reduce the amount of care sufficiently to reduce the cost of the system.

A care process must work flawlessly *end-to-end* to improve quality sufficiently to reduce the amount of care needed and the cost of the system. In other words, every point in the care process has to do its part almost every time.

Ventilator management provides an illustration. Ventilator-associated pneumonia is a common, costly ($5 billion annually for the United States), and frequently fatal complication of mechanical ventilation. Consensus guidelines recommend a bundle of several evidence-based practices to reduce the risk. The Vanderbilt University Medical Center (VUMC) matched a bundle of seven practices to the conditions of its patients and the capabilities of the health system; simplified and standardized care processes to make it easier to follow the practices; embedded the practices in order sets and documentation templates in the electronic health record system to reduce dependence on memory; and measured how consistently the practices were performed and the rate of ventilator-associated pneumonia. After these changes, performance on individual practices began to approach 90%, but all seven practices were performed concurrently for every patient just 26% of the time. Incidence of ventilator-associated pneumonia trended down, but it did not plummet.

The focus of the effort shifted to improvement of concurrent performance of all seven practices. A dashboard displayed the status of each patient's care across the practices as a row of red, yellow, and green lights. This visual cue let everyone on the care team see when performance was beginning to fall behind in time to take corrective action. Analytics helped unit managers see where their teams were getting behind by time of day, and so on. Concurrent performance began to rise. As concurrent performance exceeded 60%, the incidence of ventilator-associated pneumonia fell by about half, and both trends have continued to improve. VUMC estimates 75 deaths have been averted, 5,000 inpatient days saved, and $20 million saved over 5 years in their center alone.

The first part of VUMC's ventilator management work focused on the bundle of seven practices *individually* and improved those process measures more than it improved clinical outcomes or cost. The second part of the work focused on *concurrent performance* of all seven practices had less impact on the individual process measures, but more impact on clinical outcomes and cost.

The HITECH Act provides incentives for "meaningful use" of "certified" electronic health records. The regulations to implement these incentives focus on a healthcare provider's use of the technology for tasks at *points* in the care process such as entering prescriptions electronically, receiving decision support, transmitting prescriptions electronically to the pharmacist, making information available electronically to patients, and reporting performance on process quality. To receive incentives, providers measure and attest to performing the specified tasks electronically for a percentage of patients, or a number of clinical situations, exceeding minimums established in the regulations. Every 2–3 years, the number of tasks and minimums are advanced in "stages" of meaningful use.

The country can quickly get more value from electronic health records by shifting the focus to using the technology as part of efforts targeting *end-to-end improvement* of high-risk care processes. Drawing on the ventilator management example, each effort would include a coordinated set of measurements, management reports, care practices, decision supports, and documentation templates to support the improvement.

Start by changing the regulations to require providers to demonstrate that they *continuously* coordinate adaptation of evidence-based practices and electronic health record decision/documentation supports to improve clinical outcome and concurrent process measures.

To support this change, first require providers to report the measures they use to track progress, the care practices, and the decision/documentation supports as a "structured abstract" of an "improvement pack" to a transparent site functioning as ClinicalTrials.gov. Second, require providers to submit the "algorithms" they use to compute the measures as structured statements easily translated into programs in various electronic health records. Third, require submission of the calculated measures.

Providers can then see who is doing better and drill into their practices and supporting tools. This approach would result in a system that learns as it is used and self corrects.

RECOMMENDED READINGS

Stead WW, Lin HS, editors. Computational technology for effective health care: immediate steps and strategic directions. In: *Committee on Engaging the Computer Science Research Community in Health Care Informatics; National Research Council.* Washington, DC: National Academies Press; 2009.

Stead WW, Patel N, Starmer JM. Closing the loop in practice to assure the desired performance. Trans Am Clin Climatol Assoc 2008;119:185–195. PMID 18596845.

Stead WW, Gregg WM, Jirjis JN. Extending closed-loop control to the management of chronic disease. Trans Am Clin Climatol Assoc 2011;122:93–102. PMID 21686212.

15

BIG DATA IN HEALTH AND HEALTH-CARE: HOPES AND FEARS FOR THE FUTURE

ELIZABETH A. MCGLYNN

Big data: everyone has an opinion, experts abound, and the expressed hopes and fears are nearly as big as the data itself. At the risk of adding to the tower of Babel, this chapter offers a definition of big data, provides some examples of the type of data that are being leveraged in health, highlights some of the ways in which the big data realm is different from our historical scientific approaches, offers some warnings, and ends with an exhortation to get engaged.

Although there are numerous definitions of big data, I think the "four Vs" are a good starting point – volume, velocity, variety, and value. Volume means big. Kaiser Permanente has about 30 petabytes of data on its more than 10 million members – about the amount of data that Google processes daily (which itself is thousands of times the amount of data in the printed volumes held by the Library of Congress). Velocity means that the speed with which new data are added is increasing rapidly. By one estimate, the amount of information that is created doubles every 3 years. Variety means that the information assembled comes from disparate sources (e.g., medical records combined with social media exchanges, environmental measures, GPS tracking of places people have gone, and consumer's product-purchasing patterns). Big data is a "mashup" of different types of data. Value means that the

Perspectives on Complex Global Challenges: Education, Energy, Healthcare, Security and Resilience,
First Edition. Edited by Elisabeth Paté-Cornell, William B. Rouse, and Charles M. Vest.

insights produced from big data analyses are a reasonable basis for action and that leaders are willing to use those results to drive key business decisions.

So what are the hopes for big data in health and healthcare? A simple answer is that discovery and change can be bigger, faster, cheaper, and easier in a big data world compared with a little data world. Because health and healthcare is multidimensional, this pace of insight can be applied to enhancing business competitiveness, improving who does what and how in care delivery, accelerating scientific discovery (the "omics"), personalizing medicine, and providing insights into effective approaches to changing behavior. This is primarily driven by the idea that predictions based on correlations – the observation that two or more things are related to each other – are adequate. Big data means never having to prove causation.

All that sounds pretty good, right? So, what is the fear? For starters, people worry about spurious correlations, especially when data that do not normally travel together are combined (such as storks in Florida being predictive of birth rates). This fear in part assumes that we do not make mistakes when working in the realm of causal explanations and unfortunately that is not true. So spurious correlations will happen and those who use the results of big data should keep that reality in mind. But the big data movement is not mindless: we still need people to interpret findings and place the results in the context of other information. This means building in checks and balances. Another fear is that the data, especially in healthcare, are too "messy" and that the ways in which the data are inaccurate is not completely at random. Compare, for example, digitized data that are generated without humans entering data (information from implanted defibrillators, GPS tracking) compared with data entered by humans (medical record documentation, surveys). We definitely worry about this in "small data" analysis – and that is less problematic in "big data" because the sheer size overwhelms the dirtiness of the data. Honestly, I do not think we really know yet how much we should worry about this – but it is likely we will learn more going forward. For many companies, an issue is the storage challenges associated with the velocity with which data are being created. Many have to face questions about what to keep and what to destroy. Since we are not sure what is important, these are difficult decisions. We will likely learn over time what the appropriate shelf life is for some data (tweets vs biological information) or some technological breakthrough will make storage a nonissue (such as the recent report on storing information on a synthetic strand of DNA). Another fear is the "creepy factor" – the uses of the data will be intrusive and cause harm. That is already happening in other areas (e.g., the recent uproar over retailers using shoppers' cell phone GPS to track movements through a store) and is likely to happen in healthcare. We need to have serious societal conversations about the ethical principles that will guide this work.

Two realities of human behavior offer a bit of solace. First, embracing big data analysis as a basis for action requires that many leaders change how they make decisions. It is likely that many will ignore the insights offered by big data and continue to trust their intuition. Second, insight rarely leads directly and rapidly to action. We have plenty of examples in healthcare of knowledge that does not find its way into practice. The National Institutes of Health support significant work to unlock the secret of how to disseminate and translate insights from basic science to the bedside and have not yet solved the puzzle.

So, where does this leave us? We definitely are not going back to a nondigital age: big data is here to stay. It seems likely that big data will encounter both successes and failures along the way (this is true of all scientific inquiry). The challenge, as always, is to learn from both the successes and failures. It is also unlikely that the scientific enterprise as we know it will completely disappear. One future path is one where big data and little data analyses and methods interact – with big data pointing to smoking guns and little data testing or monitoring changes that might be made in response to those insights. So, leaders in healthcare should learn what they can about the world of big data, get involved in using its methods in their setting, and seek a balance between healthy skepticism and embracing innovation.

16

MEDICAL EDUCATION: ONE SIZE DOES NOT FIT ALL

Lloyd B. Minor and Michael M.E. Johns

The Flexner report in 1919 provided a stimulus for dramatic changes in medical education. No longer would physicians be educated through loosely organized apprenticeships with unspecified training objectives and indeterminate lengths of servitude. As a result, the curriculum in medical schools became much more standardized. The core knowledge and – to a lesser extent – skills considered necessary to practice medicine would gain widespread acceptance and would be evaluated through examinations administered to almost all medical students in the United States. While changes have occurred over time, such as the introduction of interactions with patients earlier in the course of the traditional 4-year medical school curriculum and less time passively sitting in large lecture halls, this model of medical education still exists today.

The range of activities, diversity of responsibilities, scope of postgraduate training, and sheer quantity of information and new technologies in medicine has increased exponentially over the past five decades. The changes have led to complex and evolving roles for physicians that have thus far not achieved full recognition in medical school curricula. Students continue to progress at a predetermined pace with learning objectives that vary little based upon the intended career path. There continues to be inordinate emphasis on the acquisition of didactic knowledge and time served in clinical settings and not

Perspectives on Complex Global Challenges: Education, Energy, Healthcare, Security and Resilience, First Edition. Edited by Elisabeth Paté-Cornell, William B. Rouse, and Charles M. Vest.
© 2016 John Wiley & Sons, Inc. Published 2016 by John Wiley & Sons, Inc.

enough emphasis on the development of skills required for lifelong learning and critical reasoning. The student has often been seen as an empty vessel to be filled with facts. Further, many see today's medical education as too expensive, too long, and not producing the workforce society wants in the places of need.

There are significant pressures and multifaceted challenges to better preparing health professionals to serve society in rapidly changing health-care delivery and discovery environments, and so it follows that the path to reform and realignment will be complex. One obvious point of departure is the urgent need to define core competencies that should be mastered by all physicians and to distinguish those from the more specific knowledge and skills associated with different practice-related activities linked to the various specialties of medicine. The dialogue that is needed to identify and assess these core competencies will also add clarity to the ever-changing roles of physicians. Working well as a member of teams, meeting the healthcare needs of an aging population with an increasing number of chronic diseases, addressing health disparities, and promoting quality and continuous process improvement are but a few of the needs from the physician workforce of the future. So too is the need to have a subset of physicians who are actively engaged in the translation of scientific advances to the care of patients and the development of scientific approaches inspired by disorders identified in patients. The continuous and steady advancement of medical and scientific knowledge will of necessity increase the need for specialists. But that is primarily the role of graduate medical education. Medical school education is rightly focused on creating the "stem cell" doctors – graduates ready to pursue any path in the graduate medical education (residency) phase of their education.

Advances in instructional methodologies such as online education, podcasts, new web- and app-based tools, and simulations will accelerate the development of more cohesive links between specific knowledge and skills and the different roles of physicians. Once these relationships are developed, logical and coherent educational pathways can then be tailored to meet the goals and needs of both students and society. This has the potential to create a more efficient education that could help address the impact of the growing student debt burden. Furthermore, an increasing number of medical schools may choose to focus their resources on training for a subset of the overall profiles of physician activities, such as physicians focused on the practice of primary care medicine.

The benefits that will accrue from these reforms are enormous. Reductions in the length of training for some physicians, closer interactions with schools in other health professions, and reductions in the cost of medical education are all within grasp. These instructional innovations will also have a profound impact on continuing medical education and in the creation of a culture of lifelong learning. Most importantly, the process of achieving better alignment between educational objectives and the roles of physicians will improve the ability of physicians to meet the needs of society.

Moving these curriculum changes from concept to implementation requires the commitment and leadership of medical school deans and other leaders in medical education. Such processes, of necessity, proceed incrementally because of the complex interactions of multiple constituencies and competing demands for resources. Progress is already evident at many medical schools across the country. The exchange of ideas and information among medical educators will play an important role in creation of the momentum required for success. The Liaison Committee on Medical Education, cosponsored by the Association of American Medical Colleges and the Council on Medical Education of the American Medical Association, must continue to assist in this process by encouraging innovation through its accreditation reviews.

SECTION IV

SECURITY

17

INTRODUCTION

The national security ecosystem is faced with addressing several challenges:

- Emergence of non-state powers and terrorist groups.
- Resizing the US nuclear arsenal.
- Cybersecurity.
- Intelligence.
- Biological weapons.
- US defense budget.

This introductory section outlines these challenges.

EMERGENCE OF NON-STATE POWERS AND TERRORIST GROUPS

The national security landscape has changed drastically in the last decades, especially since 9/11 and the US engagement in conflicts in Iraq and Afghanistan. Non-state entities such as Al Qaeda and ISIL operating from these areas have adopted classic and barbaric terrorist strategies and, where it is involved, the United States is confronted with an asymmetric conflict.

Perspectives on Complex Global Challenges: Education, Energy, Healthcare, Security and Resilience,
First Edition. Edited by Elisabeth Paté-Cornell, William B. Rouse, and Charles M. Vest.
© 2016 John Wiley & Sons, Inc. Published 2016 by John Wiley & Sons, Inc.

It is certainly not the first time in history, but the fight against terrorists has taken a new dimension.

The possibility that they could get their hands on nuclear weapons raises new fears and calls for new control measures. The number of nations possessing or developing nuclear arsenals is increasing. Some, such as Iran and North Korea, operate outside the general norms of international agreements, spreading nuclear technologies, and fostering nuclear proliferation among their neighbors and potential targets. Graham (2010) stated in the Nuclear Terrorism Fact Sheet of the Harvard Kennedy School that two known groups had attempted to acquire nuclear material on the black market and that a nuclear device could be delivered to its target by a number of existing means, including unauthorized immigrants and cargo ships.

RESIZING THE US NUCLEAR ARSENAL

In the last 10 years, a group of statesmen in the United States have proposed the goal of "a world free of nuclear weapons" (Shultz et al., 2007). While recognizing that deterrence still has some relevance, they argue that the costs and the threats – including those of accidental strikes – call for their gradual elimination. President Obama, in 2009, announced "America's commitment to seek the peace and security of a world without nuclear weapons." The hope is that the rest of the world, through a sequence of diplomatic agreements and moral suasion, would follow the US example.

Others argue that while the reduction of nuclear stockpiles in the United States and elsewhere is desirable, a reduction to zero – or even extremely low levels – might cause deterrence failures that could make the world even more dangerous. Schelling (2009), for instance, asks, "Why should we expect a world without nuclear weapons to be safer than one with (some) nuclear weapons?" at a time when Russia and China (among others) are modernizing their nuclear arsenals and new nuclear powers have emerged. His view is that countries that currently have a nuclear arsenal (including the United States), if they were to reduce that arsenal to zero, would have "hair-trigger" plans for the remobilization of their nuclear forces if they saw the need for it. Indeed, the thoughts have evolved in the last 10 years about the desirability of eliminating all US nuclear weapons given the realities of proliferation and other countries' expansion pressures. Along these lines, Bracken (2014), in an interview with the Bulletin of the Atomic Scientists, states that "the United States needs more than zero nuclear weapons" and that the President's

2009 Prague speech "does not seem to have convinced Russia, China, North Korea, Israel, Pakistan, India or Iran to go down the weapons-free road."

The question is thus to find an effective balance between the huge destruction potential of nuclear weapons, as Georges Shultz reminds us from his World War II experience, deterrence effectiveness, and the possibility of accidental or unintentional attacks.

CYBERSECURITY

On another front, the increasing frequency and the severity of cyberattacks have raised the fear of massive losses of personal data, operational capabilities, and industrial trade secrets. Like their colleagues in education, national security experts believe that cybersecurity is a domain that will make or break the prominence of the United States in its ability to protect its citizens and its interests. Small-scale hackers, criminal organizations, and state-level persistent attackers have targeted US government entities, industrial firms, banks, universities, and more, with a wide range of objectives, including stealing intellectual property and personal data and disrupting the functions of critical infrastructures and banks. The transparent US system is compared by Brenner (2013) to a "glass house" from which other nations have extracted, among other things, designs of US Navy systems, ATM data, and much more.

Exchange of information about cyberattacks would certainly be helpful in identifying sources and methods of cyber intrusions; but effective collaboration among industries that may compete among themselves – and with the government – is not yet fully implemented and will be difficult to arrange. Cybersecurity, however, is one domain in which public–private collaboration would be most fruitful, provided that an atmosphere of trust prevails. Adversaries, from persistent foreign attackers, to criminals and petty hackers, face little cost or consequences of their actions. Within the United States, law enforcement can be a deterrent to the extent that the perpetrators are identified and caught, which in itself is difficult enough. The problem is even more complex in cases that involve foreign entities. Retaliation against cyberattacks is mostly illegal in the United States at this time. First, there is not always a "return address" on an attack and the trajectory of cyberattacks may be complex and misleading; and second, one may want to avoid the danger of escalation. Yet, conscious of the need for protection, Obama (2015) signed an Executive Order on "Blocking the property of certain persons engaging in significant malicious cyber-enabled activities." The US decision to retaliate is thus not only technical and organizational but also legal, strategic, and political.

INTELLIGENCE

The role of intelligence based on data collection is essential in anticipating and foiling planned attacks. There, the country as well as the rest of the Western world, faces a trade-off between privacy and security. Recent revelations that the intelligence community (the NSA in particular) was gathering large amounts of information that some believe should have remained private have caused a backlash against these practices and the intelligence community at large. On the other hand, if some attacks are anticipated and prevented, these successes are not in the public domains. The Executive and Legislative powers in the United States thus face choices that affect profoundly both the privacy and the security of citizens of the United States, its allies, and its business partners.

As part of the national discussion about the balance between "the powers of government and the rights of the governed," and at the request of the Director of National Intelligence, the National Academy of Sciences (2015) provided some alternatives to the bulk collection of signals intelligence, with the goals of "collecting and/or storing less information, better protecting the information that is collected or stored against theft or compromise, and rigorously enforcing the rules governing use of collected or stored information." One of the conclusions is that "no software-based technique can fully replace the bulk collection of signals intelligence, but methods can be developed to more effectively conduct targeted collection and to control the usage of collected data." The committee recommends "protecting use rather than limiting the collection of sensitive data."

The debate on how to handle the trade-off between collecting signals and protecting privacy is international, and Herb Lin argues in this book for a role of the market in that endeavor.

BIOLOGICAL WEAPONS

In the last decades, the threat of chemical and biological attacks, especially in the hands of terrorist groups, might have come to match that of nuclear weapons, starting with anthrax attacks in 2001. Chemical weapons have been around for more than a century, for the first time on a large scale at the battle of Ypres in 1915 during World War I. In more recent times, Saddam Hussein has used them against the Kurds in 1988 in the Halabja chemical attack, killing more than 3,000 people.

One main threat now is that of bacterial and viral attacks. A country such as the United States can afford some level of preparation and storing of vaccines. But new viruses and molecules are created regularly. Keeping the results of legitimate research in that area out of the public light is a tough task, so is distinguishing natural epidemics from malicious attacks. The problem is that the threat is very diffuse, and the paradigms developed to address the treats of nuclear weapons do not apply. Research results and technologies are widely generated and communicated and the United States alone cannot control that dispersion.

A National Academy of Science report on Globalization, Biosecurity, and the Future of the Life Sciences (2006) concludes that "The global technology landscape is shifting so dramatically and rapidly that it was simply not possible […] to devise a formal risk assessment of the future threat horizon, based on the possible exploitation of dual-use technologies." A dilemma faced by the scientific community is thus the "redaction of sensitive data in the publication of dual use research of concern" (Casadevall et al., 2013).

Some biological weapons such as anthrax are targeted; others are communicative, and epidemics will not stop at borders. The diversity of threats call for vigilance regarding research, protecting the products created, manufacturing defensive products on an appropriate scale, early detection and identification of attacks and, finally, an organized response if biological weapons hit the United States.

US DEFENSE BUDGET

The United States faces a fundamental defense budget problem linked to the economics of defense, the political climate of the country, and shifting geopolitical realities. The question is: what proportion of the national budget in times of slow economic growth should be dedicated to national security, given the inertia of the military acquisition system, the rivalries among the services when it comes to budget matters and, at the same time, the need to maintain and expand US capabilities, on the ground, at sea, in the air, and in space in times of turbulences?

Sequestration in the last few years has restricted considerably the resources dedicated to defense. The framework of the Hamilton Project (Roughead, Schake & Williams, 2013) proposed "fifteen ways of decreasing the defense budget," essentially through legislation changes and personnel reduction (Roughead & Schake, 2013). Indeed, the defense budget reached a peak in 2013 and decreased both in 2014 and 2015.

For 2016, however, the Chief Financial Officer of the Pentagon states that "the geopolitical development of the last year have only reinforced the need to resource the DoD at the President's budget level rather than the current law." Citing the DoD's response to ISIL, an offensive into Iraq, the Ebola virus outbreak, and Russia's aggressive acts against its neighbors, the DoD is seeking a budget increase of about 4.4%, which Ashton Carter, the current Secretary of Defense, has called "responsible, prudent, and essential" (Carter, 2015).

The key to maintaining US competitiveness is thus in exercising wisely a leadership role, in a global world where issues of energy supply, expansion moves by Russia and China, and the development of an extremist Islamist movement have made the US national security mission much more complex. Yet, there are real dangers in inaction. The challenge is to act wisely but decisively, with a long-term strategic vision, in order to defuse what could escalate for many reasons into devastating global conflicts or massive disruptions of our societies. The question of course is to provide the DoD with what it truly needs, keeping the country safe of conventional as well as terrorist attacks, while striking a balance among, education, healthcare, energy development, and many others of the critical needs of the United States discussed here. The result of course has to reflect the political choices of the American people faced with the realities of geopolitical shifts.

OVERVIEW OF CONTRIBUTIONS

The national security landscape has changed drastically in the last decades, especially since 9/11 and the US engagement in conflicts in Iraq and Afghanistan. The fight against terrorists has taken a new dimension. The possibility that they could get their hands on nuclear weapons raises new fears and calls for new measures. At the same time, the number of nations possessing or developing nuclear arsenals is increasing. Some, such as Iran and North Korea, operate outside the general norms of international agreements.

A group of statesmen in the United States have proposed eliminating nuclear weapons altogether. Others argue that while the reduction of nuclear stockpiles in the United States and elsewhere is desirable, a reduction to zero might cause deterrence failures that could make the world even more dangerous. On another front, the increasing frequency and the severity of cyberattacks have raised the fear of massive losses of operational capabilities and industrial trade secrets. Like their colleagues in education, national security experts believe that it is a domain that will make or break the prominence of the United States in its ability to protect its citizens and its interests.

Michael E. Leiter, a former director of the US National Counter-Terrorism Center, advocates "Vigilance in an Evolving Terrorism Landscape." He observes that counterterrorism issues seem to dominate the political scene in the United States and indeed, the US psyche. On the one hand, some terrorist organizations such as Al Qa'ida have lost some of their power and cohesion, and the intelligence community has made some real progress in thwarting their destructive efforts. What level of domestic intelligence the United States wants in that respect needs to be part of a serious discussion of the trade-offs involved. Terrorists' attempts to obtain weapons of mass destruction – nuclear and biological – present a terrible threat and the huge challenge of effective resource allocation to mitigate risk of low-probability, high-consequence attacks that could wreck havoc on the country. Leiter argues that the executive branch of the US government should address this problem by continuing to deliver a nonalarmist but clear and effective message to the public. At the same time, appropriate resources should be dedicated to the monitoring and detection of weapons of mass destruction in the country's most vulnerable areas. In any case, in times of budget restriction, all US agencies and departments should work together to address counterterrorism problems and increase the US capabilities while working closely with its allies to keep the country safe.

Herbert Lin, a Chief Scientist at the National Research Council, discusses "The Market's Role in Improving Cybersecurity." He focuses on cybersecurity as a growing challenge for both the nation at large and individual entities public or private. He argues that leaving each of these entities to fend for itself behind closed doors undermines the national effort and the collective security. The nature of cyberterrorism is changing and involves a wide range of actors – hackers, criminal organizations, terrorists, and states. The spectrum of cyberattacks has increased widely and now threatens the country's vital infrastructure, including the electric power grid, the air traffic control system, communication networks, and the banking system.

Lin considers the fragmentation of the response to these threats as a real obstacle to appropriate investment in cyber defense. He points to several causes of this deficiency: there is no real incentive to share information about cyberattacks, the costs of cybersecurity countermeasures are often overestimated, and markets may allow some potential targets to escape the costs that they should pay for what is in reality, an economic externality. In other terms, each cares about its own protection but does not see as its responsibility the protection of others as part of a global system. To address this economic and behavioral challenge, Lin proposes several measures.

A starting point is to make these entities understand the depth, scope, and dependencies of cyberthreats as well as the benefits of past security investments, and to require that they report all breaches of their system. He also proposes to put in place economic and judicial procedures that would make system vendors and operators liable for the damage caused by breaches involving their system in any way. He also advocates regulations and certification of conformance for operators of critical infrastructure. He thus describes a regulated market in which all actors share the costs, the benefits, and the responsibility of the country's cybersecurity.

George P. Shultz, former Secretary of State, provides his perspective in "On Nuclear Weapons." He is a firm advocate of the reduction of the US nuclear stockpile to zero, with the understanding that this would have to involve an agreement of other countries to pursue the same goal. He presents the case by recounting the story of his experience as a young marine at the end of World War II and the horror that he felt when he realized the effects of nuclear bombs in Japan. As Secretary of State under President Regan, he was an active participant in the discussions that led to the Reykjavik agreement. Ronald Regan and Michael Gorbachev agreed at that time that the threat of nuclear weapons had to be eliminated by eliminating the weapons themselves. Shultz describes the hostility of some of the US allies, the British in particular, to that idea, but also the success in decreasing the total number of nuclear weapons since the Reykjavik talks. He acknowledges the immense difficulties in achieving a zero goal. If nothing else, countries such as North Korea are developing them actively, and India and Pakistan keep each other in check by the presence of their own arsenal. Nonetheless, Shultz thinks that it is essential to work immediately toward a general agreement on the objective to eliminate nuclear weapons and a path to reaching that goal.

Siegfried Hecker, a former director of Los Alamos National Laboratory and a professor of engineering, provides his views in "The Nuclear Security Challenge: It's International." He links the threats of nuclear weapons to the growth of the nuclear power sector worldwide in what he calls the "dual-use" dilemma. A large number of nuclear reactors are under construction worldwide with the promise of abundant, clean, and cheap energy. At the same time, nuclear wastes provide the raw material for nuclear weapons. During the cold war, the threat of mutually assured destruction was a bilateral problem, but the worldwide spread of nuclear technology has made the problem almost intractable and the nuclear nonproliferation regime less and less effective. Various sources of enriched fuel through nuclear energy generation, outright theft, or acquisition of enrichment systems (e.g., from the AQ Kahn network out of Pakistan) raise the threat of possession and use of these weapons by terrorist groups and rogue nations. Hecker acknowledges the immense complexities – political, military, diplomatic, and technical – of addressing the proliferation problem and he concludes that nothing can be effectively done unless the major powers that openly own nuclear weapons at this time agree to reduce their nuclear arsenals and their reliance on nuclear weapons.

Henry Kissinger, former Secretary of State, and Brent Scowcroft, a retired US Air Force Lieutenant General and US National Security Advisor, argue that "Nuclear Weapon Reductions Must Be Part of Strategic Analysis." They also address the problem of decreasing the size the US nuclear stockpile and discuss the end point of the reduction process. They place the required negotiations in the context of global peace and a strategic analysis of US security needs. While supporting the objective of decreasing the size of the US and others' nuclear stockpiles, they argue that very low numbers – zero *a fortiori* – could create a dangerous situation because of the limited response options in case of an attack. In any case, they recommend that no goal be stated in terms of number of weapons. Instead, they put the emphasis on nuclear capability, including delivery vehicles and weapons types. Whereas they agree that Russia is still the most powerful adversary in nuclear terms, they call the country's attention to other kinds of adversaries, their potential alliances, and their values, which might undoubtedly be very different from those of the United States. In any case, they stress the importance of the trust of our allies in the existence and effectiveness of the American protection. They recommend enlarging the diplomatic negotiations of international non-proliferation treaties, while maintaining a strategic and tactical force sufficient for credible deterrence to a first strike.

Jacques S. Gansler, a professor of public policy and former Under Secretary of Defense, addresses "Maintaining US Leadership in Science, Technology, and Innovation for National Security." He focuses on the importance of leadership in science and technology to maintain the strategic advantage of the United States. He too is concerned by the weakening of the US education system and the loss of US students' interest in science and technology. He is also concerned by recent declining trends in the US defense industry and its research investments. The United States like the rest of the world is increasingly part of a global system. In scientific domains, US universities rely in large part on foreign graduate students and the United States has been a net importer of high technology in the last decade. For example, it has to import critical military technologies at a time when they are much needed. To maintain a sustainable level of independence, Gansler emphasizes the need for investments not only in technology but also in processes designed to supply better and lower-costs equipment and services. He advocates an increase in international R&D, retention of some of the best foreign students trained in the United States, but also stabilization of the DoD budget dedicated to the development of "disruptive technologies." His general conclusion is that the United States has to embrace the realities of globalization, and come to grips with military budget restrictions while reacting swiftly to technological changes that affect its security.

REFERENCES

Bracken P. *Interview: Paul Bracken on American Nuclear Forces in the 21st Century.* Bulletin of the Atomic Scientists; 2014. December..

Brenner J. America the vulnerable: the new threat matrix of digital espionage, crime and warfare. In: Brenner J, editor. *Glass Houses: Privacy, Secrecy and Cyber Insecurity in a Transparent World.* New York: Penguin; 2013.

Carter A. Proposed budget is responsible, prudent, and essential. Testimony to Congress; 2015 Mar.

Casadevall A, Enquist L, Imperiale MJ, Keim P, Osterholm MT, Relman DA. Redaction of sensitive data in the publication of dual use research of concern. mBio 2013;5(1):e00991–e010013.

Graham A. *Nuclear Terrorism Fact Sheet.* Cambridge, MA: Belfer Center, Harvard Kennedy School; 2010. , Apr.

National Academy of Sciences. *Globalization, Biosecurity, and the Future of the Life Sciences.* Washington, DC: National Academy Press; 2006.

National Academy of Sciences. *Bulk Collection of Signals Intelligence: Technical Options.* Washington DC: National Academy Press and Office of the Director of National Intelligence; 2015.

Obama B. *Remarks in Prague.* Washington, DC: The White House, Office of the Press Secretary; 2009. , Apr 5.

Obama B. *Blocking the Property of Certain Persons Engaging in Significant Malicious Cyber-Enabled Activities*Executive Order.. Washington, DC: The White House; 2015. , Apr.

Roughead G, Schake K. *Hamilton Project. National Defense in a Time of Change.* Washington, DC: Brookings Institute; 2013.

Roughead G, Schake K, Williams C. *Hamilton Project. Real Specifics: 15 Ways to Rethink the Federal Budget. Part 1: Budgeting for a Modern Military.* Washington, DC: Brookings Institute; 2013. , Feb 22.

Schelling TA. A world without nuclear weapons? Daedalus 2009;138(4): 124–129Fall.

Shultz G, Kissinger H, Perry W, Nunn S. A world free of nuclear weapons. The Wall Street Journal; 2007 Jan 4.

18

VIGILANCE IN AN EVOLVING TERRORISM LANDSCAPE

Michael E. Leiter

To say that terrorism has been front and center in national security discussions over the past 15 years would indeed be an understatement. Since September 11, 2001, issues related to terrorism have received considerable attention in our elections, our national and homeland security institutions, and perhaps all-too-often the public's psyche. The tenor of this discourse has modulated over time: the death of Osama bin Laden and the tenth anniversary of 9/11 brought forth optimism and even some premature predictions of victory. The rise of the Islamic State of Iraq and Syria (ISIS), along with a troubling spate of successful terrorist attacks in Western Europe and the United States, quickly reversed this trend and brought with it even more dire and apocalyptic prophecies. The challenges of terrorism today share both similarities and differences with that which we faced in 2001. As importantly, however, the significant and extremely rapid fluctuations in how we perceive the state of terrorism poses a serious challenge to our ability to address appropriately the full range of terrorist threats in coming years.

Looking back: winning in 2012. Although it is always difficult (and politically hazardous) to estimate accurately and precisely the state of the terrorist threat, the vast majority of serious observers agree that in 2012 the risk of a large-scale, catastrophic terrorist attack in the United States was the lowest it had been since 9/11. From my perspective, this was certainly the case: al

Perspectives on Complex Global Challenges: Education, Energy, Healthcare, Security and Resilience,
First Edition. Edited by Elisabeth Paté-Cornell, William B. Rouse, and Charles M. Vest.
© 2016 John Wiley & Sons, Inc. Published 2016 by John Wiley & Sons, Inc.

Qa'ida in Pakistan was a shadow of its former self – and unquestionably so with the death of bin Laden. Al Qa'ida's next most potent affiliate, al Qa'ida in the Arabian Peninsula based in Yemen, certainly posed a threat to the region and the United States (as illustrated by attacks such as the Christmas Day bomber and its attempt to detonate bombs on two cargo flights en route to the United States), but even this threat was less deadly and less broad than that which we faced in 2001.

Looking elsewhere in the world, the al Qa'ida threat was even less orga- nized and likely less able to pull off an attack anything close to the tragedy that killed almost 3,000 in September 2001. Whether it was North Africa, East Africa, Southeast Asia, or Europe, al Qa'ida affiliates were – even where experiencing greater freedom to organize than in the past due to events asso- ciated with the early days of the Arab Awakening – far from the methodical and efficient al Qa'ida that we had faced in the past. This was equally if not more true of the colloquially termed "homegrown threat": US residents who have either spontaneously or with guidance from al Qa'ida elements over- seas sought to attack in the homeland in al Qa'ida's name. And although al Qa'ida's ideology has not disappeared, its pull both overseas and at home was at best flat and more likely in decline.

There is perhaps no illustration of this counterterrorism success roughly a decade after 9/11 more powerful than the truly remarkable fact that since 2001, al Qa'ida and those it inspired had killed only 14 people in the United States. Although every loss is tragic, we must also recognize that it in terms of scale this is the same loss of life as the number of teenagers killed in the United States *every day* in automobile accidents. In short, we were, I believe, undoubtedly improving the situation.

Of course even in these heady days of 2012, this was the good news but not the whole story. Good intelligence, offensive successes, able partners, and defensive measures had made terrorists' work more difficult but their efforts continued. One can easily imagine how different (and less optimistic) this story would have been had Northwest Flight 253 been downed over Detroit on Christmas Day 2009, if multiple cargo jets disappeared from radar in 2010, or scores had been killed by the failed Times Square bomber that same year.

Fast forward: losing in 2016? A few short years later and how the picture has changed. As governments fell or faced massive upheaval across North Africa and the Middle East, ISIS picked up al Qa'ida's global mantle, claimed to have established a caliphate across Syria and Iraq, and launched affiliates across the globe. Simultaneously, ISIS advanced on the ideological battlefield by mastering widely accessible Internet-based tools to propagate its message, recruit new adherents, and communicate with increasing security.

The result can be measured in many ways but none are particularly positive for those opposing ISIS. More Western foreign fighters joined ISIS's ranks than ever before seen with al Qa'ida or its affiliates. More wealth is controlled in its caliphate than could be contemplated during al Qa'ida's heyday in Afghanistan, Pakistan, or Yemen. And although ISIS's attacks in the West have not remotely approached 9/11's tragic scale, the group's ability to inspire adherents or direct members to launch attacks in Brussels, Paris, California, and elsewhere represent a degree of traction with certain communities that al Qa'ida never achieved.

Consequently, the urgent cries of the West facing a rising and existential threat have reemerged. Political effects are both real and jarring: calls for significant changes to the European Union, tensions between governments and technology companies over access to information, and proposals to tighten borders based on religious litmus tests.

Finding meaning for the future. What does this significant swing, over less than 4 years, teach us for the future? In short, it highlights the importance of avoiding short-term decision making, prioritizing ongoing threats with potentially cataclysmic consequences, growing counterterrorism partnerships, and rationalizing budgetary choices. None are easy – especially in a hyper-partisan environment – but all are necessary to optimize our likelihood of addressing terrorism effectively.

The terrorism pendulum. These 4 years have clearly illustrated how major fluctuations in public and political perception of terrorism risks can lead to questionable and short-termed decision making. In 2012, after 10-plus years of near-constant public discussion of terrorism – in our politics, the media, and elsewhere – people and politicians had simply had enough. During this period, and rapidly accelerated by disclosures by the likes of Edward Snowden, the question du jour moved quickly from "how can the Government protect us against terrorism" to "why is the Government invading our privacy?" By 2016, many of these trends reversed and there were again calls for increased surveillance and broader counterterrorism operations.

These fluctuations may be inevitable and to some extent – especially when prompting serious reflection as opposed to demagoguery – critically important to help balance potentially competing societal interests. On the other hand, we must recognize the real risk posed by such swings, for they can undermine long-term counterterrorism success in a variety of ways. For example, because many counterterrorism capabilities will take years to build and mature, political and funding consistency is critical to maintaining effective security efforts. This is even more true in working with international partners, with whom building capabilities normally takes years and who rightfully expect long-term commitments in sensitive areas such as intelligence and security.

These extreme swings in public perception also pose concrete risks to our most valuable resource: counterterrorism professionals. Specifically, such swings risk alienating security specialists who fear that rapidly shifting winds will make yesterday's national security priority tomorrow's national embarrassment. Or, going in the other direction, today's careful discretion will be viewed as tomorrow's contributing factor to a counterterrorism failure. We must of course maintain individual and institutional accountability, and find balance between intelligence gathering and respect of privacy. We must also, however, recognize that excessive swings of the terrorism pendulum can – if not carefully managed – undermine our ability to recruit and train the very individuals we need most to be effective.

Weapons of mass destruction. There is no doubt that terrorist attacks or at least attempts will continue to occur at home and abroad. Such attacks can cause enormous pain and suffering to victims and their families, but they are clearly of a scale that is dwarfed by other societal ills such as routine criminal activity. The same cannot be said of terrorists' use of weapons of mass destruction – and more specifically biological weapons or an improvised nuclear device (dirty bomb, stolen bomb, or bomb made from stolen material).

Although we have also made progress in reducing the likelihood of terrorists obtaining WMD, for the foreseeable future we are faced with the possibility that a terrorist organization will successfully acquire these weapons. In this case, technology is not yet our friend, as the ease with which these weapons can be obtained and hidden continues to exceed our ability to detect them. In no case is this truer than for biological weapons, where dual-use technology and expertise have continued to make access easier for more and more of the population.

Weapons of mass destruction pose a unique challenge, as they are the prototypical low-likelihood, high-consequence event and thus determining the proper allocation of resources to combat them is particularly contentious. That being said, we must continue to protect against the most dangerous of materials (e.g., highly enriched uranium, smallpox virus) being obtained by terrorists, secure weapons in the most dangerous places (e.g., Pakistan), and pursue research and development that will assist in detecting chemical and biological weapons in places where they would do the most harm.

Counterterrorism partnerships. Counterterrorism has always been and continues to be a "team sport." Although the United States can do much alone, we have always been extremely reliant on a vast network of friendly nations that have extended massively our intelligence, law enforcement, military, and homeland security reach. Even before the Arab Awakening, we witnessed some weakening of these partnerships. Whether it was fatigue on our partners' part, their own resource challenges, or differing views on the proper scope of counterterrorist efforts (e.g., fights over data sharing between the United States and the European Union), these partnerships have been under increasing pressure. Post-Arab Awakening, we face an even more daunting task, having lost many of our most valuable partners in the very places we need them the most.

Again, part of the challenge is that we have been a victim of our own success. Al Qa'ida's near defeat in 2012 led many to diminish cooperation and prioritize other domestic or international initiatives. It was this very diminishment that played a contributing role in allowing ISIS to recruit, train, and in some cases direct operatives in countries with increasingly under resourced and politically disfavored security services. With this came ISIS's metastasizing and embedding in ways that made them more dangerous and more difficult to displace.

To maintain our partnerships (and our partners' focus), we must carefully preserve the funding for programs that provide critical capabilities – and potentially more important, a positive US presence – for our allies. The increase in funding for special operations forces is a good step, but relatively tiny investments in the Department of State and Justice programs can also deliver real results in this realm. In addition, we must coordinate with Europe's intelligence and law enforcement apparatus so that they may be as integrated and mature as the many other, non-security elements of the European Union have become over the past quarter decade.

Resources. Finally, and not entirely inappropriately, counterterrorism budgets at the federal, state, and local levels have also fluctuated during this period. At its peak, the United States likely spent upwards of $100 billion a year on counterterrorism, and by 2012 Congress and others were looking for the equivalent of a "Peace Dividend." With ISIS's rise, this frugality was largely eliminated once again although still not to the extent of the massive, post-9/11 increases.

The question, however, is whether we will be willing or able to make smart choices to preserve critical capabilities – whether budgets are going up or down. Our historic ability to direct funds where the threat is greatest (as opposed to where the political forces are strongest) has not been good. We generally increase capabilities when budgets go up and vice versa, but it is not at all clear whether we are having a corresponding effect on the overall terrorism risk that we face. Simply said, we should not spend a marginal dollar on, say, improving Transportation Security Administration (TSA) capabilities unless this is also the best way to produce the biggest *overall* reduction in terrorism risk.

Doing this admittedly challenging analysis requires that we transition to serious mission-based – as opposed to department-and-agency-specific-based – budgeting in the federal government. This approach will require enormous changes within the executive and congressional branches, but looking across the counterterrorism budget, identifying the critical capabilities we must preserve, and then figuring out how that matches department-specific budgets can be done. And if we are serious about maintaining those capabilities that will in turn have the greatest effect on reducing risk, we have little choice.

There are no easy fixes to any of these challenges but all these risks can – similar to the terror threat itself – be reduced. As so many said immediately after 9/11, terrorism is a "generational challenge." We must not forget this during the good or bad times, for when we do our short-term choices run the risk of deepening our valleys and delaying our victories. We must have patience and focus, for this is indeed a counterterrorism marathon, not a sprint.

19

THE MARKET'S ROLE IN IMPROVING CYBERSECURITY

Herbert Lin

Cybersecurity has become an issue of pressing national importance. Many elements of cyberspace are highly vulnerable to an expanding range of attacks by a spectrum of hackers, criminals, terrorists, and state actors. For example, Government agencies and private sector companies both large and small suffer from cyber thefts of sensitive information, cyber-vandalism (e.g., defacing of Web sites), and denial-of-service attacks. The nation's critical infrastructure, such as the electric power grid, air traffic control system, financial systems, and communication networks, depends extensively on information technology (IT) for its operation. Concerns about the vulnerability of this infrastructure have heightened in the security-conscious environment after the September 11, 2001, attacks. In short, the cybersecurity posture of the nation – by which is meant the extent to which the information technology systems and networks of the nation are resistant to hostile actions conducted against them – is inadequate, and to judge from the expressions of public concern, woefully inadequate.

The US government has not been blind to concerns about the nation's cybersecurity posture, and has been at the forefront of calling attention to them since the late 1990s, and has made substantial investments to develop better security technologies. Recently, policy makers have become increasingly concerned that adversaries backed by substantial resources

Perspectives on Complex Global Challenges: Education, Energy, Healthcare, Security and Resilience, First Edition. Edited by Elisabeth Paté-Cornell, William B. Rouse, and Charles M. Vest.

will attempt to exploit the cyber-vulnerabilities in the critical infrastructure, thereby inflicting substantial harm on the nation.

With this outpouring of public concern regarding cybersecurity, why has there been so little progress in improving the actual cybersecurity posture of the nation? A 2007 NRC report pointed to three possible (and complementary) reasons (National Research Council, 2007).

1. The unavailability of information detailed enough to inspire decision makers in the private sector to take action. Along these lines, Senator Sheldon Whitehouse has written that "if the public knew the stakes – that cybercriminals have pulled off bank heists that make Willie Sutton look like a petty thief – they would demand swift action (Sheldon Whitehouse, 2011)."

2. The overweighting of the immediate costs of putting into place adequate cybersecurity measures relative to the potential future benefits (more precisely, the avoided costs) of preventing cyber-disaster in the future – and the systematic discounting of the latter as uncertain and vague.

3. Market externalities that allow the relevant decision makers to escape the full costs of inadequate action. To the extent that private parties do respond to security issues, they generally do so as one of the risks of doing business. But they care less about the risks to third parties that arise as a result of their behavior, and they do much less to respond to the threat of low-probability, high-impact (i.e., catastrophic) threats, even though all of society has a significant stake in their actions.

The picture described here suggests that the availability of appropriate information technologies and operating practices for security is only one aspect of developing a more robust cybersecurity posture for the nation. The existence of these security technologies and practices is necessary but not sufficient – they must be deployed and used on a scale commensurate with plausible attacks, which requires that sufficient incentives be in place for such deployment and use.

Taken together, these three points suggest that even if organizations and individuals pay adequate attention to cybersecurity for their own needs (and they often fail to do even that), the overall national cybersecurity posture that results from many such individual investment decisions may be much lower than the nation needs as a whole. In short, the free market has simply failed to date to deliver a sufficiently robust cybersecurity posture for the nation.

One key reason is that cybersecurity is to a large extent a public good; much of the payoff from security investments may be captured by society rather than directly by any individual firm that invests. For example, it often happens that an attacker A who wishes to attack victim V will route the attack through computer facilities owned by intermediary N, who knows neither A nor V. That is, A will compromise N's computers in order to attack V. However, the compromise on N's computers will not damage them very much, and indeed N may not even notice that its computers have been compromised. Investments made by N to protect its computers will not benefit N, but will rather protect V. But an Internet-using society would clearly benefit if all of the potential intermediaries in the society made such investments. Making parties liable for not securing their facilities against being illicitly used as part of attacks using intermediaries would change the incentives for making such investments.

A further contributing factor is that many of today's systems are generally built out of commercial off-the-shelf components. Such components are not very secure in part because of the market demand for secure components: many customers buy features and performance rather than security, and customer appreciation of system quality focuses on features that affect normal operations – speed, ease of use, functionality, and so on. The failure of the US government's Orange Book program even within the federal marketplace is a striking example. "The Orange Book" is the nickname for the US government-promulgated Trusted Computer System Evaluation Criteria, which were developed in 1983 and were intended to guide commercial system production generally and thereby improve the security of systems in use by government. The IT industry did manufacture compliant systems, but government agencies refused to buy them because they were slower, later to market, and less functional than other less secure systems available on the open market. For example, in some cases, a certified system was unable to connect to a network, because a network connection was not part of the certified configuration.

Finally, it is generally true that installing security safeguards has negative aspects such as added cost, diminished performance (e.g., slower response times), inconvenience in use, and the awkwardness of monitoring and enforcement, not to mention objections from the work force to any of the above. Against this backdrop, a number of approaches have been suggested that may change the environment in which decisions regarding cybersecurity are made. These include but are not limited to

- Liability of software and system vendors and system operators to potential damages suffered not just by customers but also by third parties for system breaches.
- Mandated public reporting of security breaches, especially those that could threaten critical societal functions or that could affect an entities'

reputation or the willingness of customers to engage with entities affected by such breaches.

- Regulation that imposes best practices on system operators of critical infrastructure under threat of civil penalty.
- Accounting practices that force companies to report their cybersecurity practices and postures and the (sanitized) results of independently conducted red team attacks.
- Certification of conformance to relevant cybersecurity standards that can be used as a marketplace differentiator.

Policy making with respect to any of these possibilities is further complicated by the question of scope – to which entities (or classes of entities) and under what circumstances would any of these policy instruments apply. The cybersecurity problems faced by entities that are part of the nation's critical infrastructure are different from those facing individual users, for example, and it is likely that different approaches to each will be more effective than one standardized approach.

Recognizing these points, the US Comprehensive National Cybersecurity Initiative (CNCI) identified cyber-economic incentives (e.g., providing incentives to good security, science-based understanding of markets, decision-making and investment motivation) as one of the "leap-ahead" technologies that would help to achieve dramatic improvements in cybersecurity. A 2011 Department of Commerce "green paper" recommended the development of incentives to combat cybersecurity threats, such as those that might reduce "cyberinsurance" premiums for companies that adopt best practices for cybersecurity and openly share details about cyberattacks for the benefit of other businesses (Department of Commerce, 2011). This paper also pointed to the need for improving public understanding of cybersecurity vulnerabilities through education and research.

The current US government view of cybersecurity recognizes the possibility that different sectors in the private and public spheres might require different approaches to improve their cybersecurity postures. In February 2014, the National Institute for Standards and Technology released its *Framework for Improving Critical Infrastructure Cybersecurity*,[1] a document intended to help critical infrastructure providers in the private sector manage cyber risks by providing a common language of standards, guidelines, and practices. No legislative or regulatory measures mandate the use of the framework, thus allowing individual organizations to manage cyber risks in a cost-effective manner based on their own assessment of their business needs.

[1] http://www.nist.gov/cyberframework/upload/cybersecurity-framework-021214.pdf.

Presidential Policy Directive 21, issued in February 2013, lays out the federal responsibility for critical infrastructure of the United States.[2] The Department of Homeland Security has primary federal responsibility for 8 out of 16 critical infrastructure sectors, specifically chemical; commercial facilities; communications; critical manufacturing; dams; emergency services; information technology; and nuclear reactors, materials, and waste. It shares responsibility with the Department of Transportation for the transportation systems sector and with the General Services Administration for the government facilities sector. The Department of Defense has primary responsibility for the defense industrial base, and the Department of Energy for the energy sector. The Department of the Treasury has primary responsibility for the financial services sector. The Department of Health and Human Services has primary responsibility for the healthcare and public health sector, and shares responsibility for the food and agriculture sector with the Department of Agriculture. The Environmental Protection Agency has primary responsibility for the water and wastewater systems sector.

Both public and private sector initiatives have been undertaken to increase the incentives for cybersecurity. For example, the National Strategy for Trusted Identities in Cyberspace (NSTIC) promotes a public–private partnership to create the standards and policies needed for interoperable trusted identification credentials that could be used online in reducing ID theft and fraud online.[3] Competitive markets have emerged for antispam services and products[4] and for antiphishing functionality in browser software and ISP services (Sheng et al., 2009). The Anti-Phishing Working Group is an international nonprofit pan-industrial and law enforcement group that focuses on eliminating the fraud, crime, and identity theft that result from phishing, pharming, malware, and e-mail spoofing.[5]

It still remains the case, however, that after many years of debate about how policy makers should proceed (or indeed if they should do anything at all), there is no national consensus about what should be done to promote cybersecurity comprehensively. Indeed, any change in the environment for decision-making will be met with suspicion by those who would be affected by such changes – not surprisingly, since their basis for business planning would, by definition, be changed.

---

[2]https://www.whitehouse.gov/the-press-office/2013/02/12/presidential-policy-directive-critical-infrastructure-security-and-resil.
[3]http://www.whitehouse.gov/sites/default/files/rss_viewer/NSTICstrategy_041511.pdf.
[4]For example, the ICSA Labs Anti-Spam Product Developers Consortium (ASPMPD) is comprised of antispam vendors of solutions that deliver legitimate e-mail messages without unwanted spam. See https://www.icsalabs.com/technology-program/anti-spam/consortium.
[5]http://www.antiphishing.org/.

How might a consensus on action be achieved? In today's hypercharged political environment, reaching consensus on anything significant is difficult. But surely a first step would be to better understand the nature and extent of systemic decision-making failures (if any) to promote cybersecurity, the nature of the "on-the-ground" decision-making environment and how that environment affects key decisions about cybersecurity, and past successes and failures of individual solutions to different aspects of the cybersecurity problem. Such understanding might help to develop principles regarding economic and other mechanisms that would change the decision-making calculus of private-sector decision makers in ways that promote a more robust national cybersecurity posture – which is a goal shared across partisan divides.

REFERENCES

Department of Commerce. Cybersecurity, innovation and the Internet economy; 2011 Jun.

National Research Council. *Towards a Safer and More Secure Cyberspace*. Washington, DC: National Academies Press; 2007.

Sheldon Whitehouse. Cybersecurity needs complete plan, Politico; 2011 Mar 7, 04:31 AM. EDT.

Sheng S. et al., An empirical analysis of phishing blacklists. CEAS 2009 – Sixth Conference on Email and Anti-Spam; 2009 Jul 16–17; Mountain View, California, USA. Available at http://ceas.cc/2009/papers/ceas2009-paper-32.pdf; Anti-Phishing Best Practices for ISPs and Mailbox Providers. Available at http://www.antiphishing.org/reports/bestpracticesforisps.pdf.

20

ON NUCLEAR WEAPONS

George P. Shultz

My experience with nuclear weapons started in an odd way. As a marine who had been out fighting World War II in the Pacific for two and a half years, I had just boarded a ship full of marines headed for the States. We knew we would be formed into the units that would assault the Japanese homelands, so we were well aware of what lay ahead for us. We were hardly out of port when we heard that something called an atomic bomb had been dropped. No one had the slightest idea what that was, but we assumed it was important because it was reported. The ship lumbered on and we heard that a second bomb had been dropped. By the time we made port in San Diego, the war was over. Whatever the atomic bomb was, we assumed it had saved a few of our lives.

I went on to study economics at MIT. I remember when I first saw the picture of Hiroshima and was appalled. We all observed nuclear tests in the atmosphere. Then, we learned how insane it was to release all that radiation. Something was wrong.

I continued my work on economics and economic policy, including time in the cabinet as Secretary of Labor, Secretary of the Treasury, and Director of the Office of Management and Budget. Nuclear issues were not on my plate, but the cloud of this potential catastrophe, should deterrence fail, was always present.

Perspectives on Complex Global Challenges: Education, Energy, Healthcare, Security and Resilience, First Edition. Edited by Elisabeth Paté-Cornell, William B. Rouse, and Charles M. Vest.

As Secretary of State, I became directly involved in negotiations to reduce the numbers of nuclear weapons. President Reagan asked the Joint Chiefs to give him an estimate of the impact on the United States of an all-out Soviet nuclear attack. Their answer: initial casualties of around 150 million and continuing casualties because infrastructure would be destroyed. Would he retaliate? Yes. I heard him say many times, "What's so good about keeping the peace by having the ability to wipe each other out?" He also said early on and very publicly that we should do everything we can to eliminate nuclear weapons from the earth.

I sat beside him in Geneva in 1985 when he and General Secretary Gorbachev agreed that a nuclear war can never be won and must never be fought and in Reykjavik in 1986 when they agreed on the objective of eliminating nuclear weapons.

That idea was greeted with hostility at the time. Prime Minister Margaret Thatcher summoned me to the British ambassador's residence in Washington and "handbagged" me. The beating took the form of asking me how I could sit there and allow the president to agree to the elimination of nuclear weapons. When I told her I agreed with the president, the upbraiding intensified. I am still deeply convinced that President Reagan was right and that we must persevere in seeking the end of nuclear weapons. As Henry Kissinger once put it, we have "stolen the fire from the gods; can we confine it to peaceful purposes before it consumes us?"

Is progress possible? Absolutely. The number of nuclear weapons in the world today is roughly one-third of what it was at the time of the Reykjavik summit. I and others have written about the vision of a world free of nuclear weapons and the steps needed to get there. Any of these steps would, in itself, make the world safer. Four important conferences attended by around 40 heads of government have taken place to seek agreement on how to get better control of fissile material – an important step. Right now we see North Korea with a nuclear weapon and the clear aspirations, though now delayed, of Iran to become a nuclear power. Those are steps in the wrong direction. We see the tension of nuclear-armed India and Pakistan with all the dangers of an escalation into a nuclear exchange.

We need a commitment to a joint enterprise where countries take responsibility for movement toward the conditions for a world free of nuclear weapons. The meetings on fissile material provide an example of what can be done.

The threat is real. The problems are immense. The path to a better future is clear. The time for positive effort is now.

21

THE NUCLEAR SECURITY CHALLENGE: IT IS INTERNATIONAL

SIEGFRIED S. HECKER

Nuclear energy can electrify the world or it can destroy it. Nuclear power holds the promise of a sustainable, affordable, carbon-friendly source of energy available this century on a scale that can help meet the world's growing need for energy and slow the pace of global climate change. Yet, the same factor of millions increase in energy density obtained by splitting the atomic nucleus compared with other energy sources that makes nuclear energy so attractive for the generation of electricity can also be used to create nuclear explosives of unprecedented lethality.

For nearly 50 years, Washington and Moscow held the fate of the world in their hands through the Cold War's policy of mutually assured destruction. The collapse of the Soviet Union made the specter of nuclear apocalypse fade to a distant memory, yet the likelihood of a nuclear attack has increased because dictatorial leaders of some failed or failing states and some stateless terrorists find that factor of millions irresistible. The new threats are limited nuclear war, the proliferation of nuclear weapons, and nuclear terrorism. An unprecedented level of international cooperation will be required to address these threats and avoid a nuclear catastrophe.

The dual-use dilemma. Nuclear energy plays so vital a role in the economic welfare and security of nations that incentives to pursue nuclear capabilities are overwhelming, while technologies for peaceful and military use are

Perspectives on Complex Global Challenges: Education, Energy, Healthcare, Security and Resilience,
First Edition. Edited by Elisabeth Paté-Cornell, William B. Rouse, and Charles M. Vest.

for the most part interdependent and interchangeable. Moreover, the symbolic significance of nuclear status today matches or exceeds its economic or military impact. The nuclear nonproliferation regime – a fabric of treaties, bilateral and multilateral agreements, organizations, and inspections designed to halt the spread of nuclear weapons while providing access to peaceful uses of atomic energy – has helped to limit the number of states with nuclear weapons. However, it is generally agreed to be under severe stress today.

The once-frightening dilemma of two nuclear scorpions in a bottle has been replaced by a more complicated international dynamic in which nuclear weapons may be sought by states that have lost their superpower protection or have regional expansionist visions. Iraq, Libya, North Korea, and Iran mounted significant nuclear weapons efforts during the past 25 years. India and Pakistan declared themselves to be nuclear weapon states by conducting nuclear tests in 1998. Israel's policy of nuclear ambiguity or opaqueness complicates the security dynamic in the Middle East. South Asia presents the greatest risk of nuclear war. Continued political turmoil is exacerbated by increased reliance on nuclear weapons. North Korea and Iran's pursuit of nuclear weapon capabilities, under the umbrella of peaceful nuclear energy, has the world on edge, particularly because their governments are dictatorial and the leadership unpredictable.

Supply and demand: Although only 10 countries are believed to possess or pursue nuclear weapons today, less than in the 1960s, the potential for global disruption is greater because the world is economically and politically so interconnected. Technical capabilities to build the bomb have become more accessible; the basic knowledge required is only a click away, and pursuit of civilian nuclear programs brings required materials and technologies dangerously close to weapons utility.

There is a long history of international cooperation to restrict supply of the knowledge, materials and technologies for nuclear weapons through International organizations and arrangements such as the International Atomic Energy Agency and the Nuclear Suppliers Group. Their success has been countered by proliferation networks such as that run by Pakistan's A.Q. Kahn aided by greedy entrepreneurs around the world. Although continued vigilance is required by the international community to restrict supply, North Korea's nuclear weapons and Iran's relentless pursuit of the nuclear weapon option demonstrates that if demand is high, countries with modest technical capabilities can work around restrictions and build the bomb.

Much more must be done to restrict demand for nuclear weapons. Countries pursue nuclear weapons primarily for national security reasons, but domestic and international considerations also come into play. The international community has continued to pursue supply-side restrictions, backed by sanctions, to counter nuclear threats from the likes of North Korea and Iran. Little has been done to address why they want the bomb. In the case of North Korea, once Pyongyang tested and declared itself a nuclear power, it used the nuclear program for domestic and international leverage.

Dealing with other nations' nuclear weapon aspirations will require that the big powers look beyond their immediate national security concerns. For example, it is difficult to convince countries not to pursue nuclear weapons unless the five accepted nuclear powers, particularly the United States and Russia, do more to reduce their nuclear arsenals and decrease their reliance on nuclear weapons. The examples of North Korea and Iran also demonstrate that scaling back nuclear programs requires that the big powers compromise to find common ground, even if it requires them individually to take more risk, both militarily and politically.

It takes so little and there is so much. Many consider nuclear terrorism the most serious global security threat. The primary concern is the acquisition of fissile materials, either by theft or diversion, by subnational or terrorist groups leading to an improvised nuclear device. Even an imperfect nuclear device of a few kilotons or less detonated in one of the world's megacities could kill 10,000 or more people and cause global disruption on an unprecedented scale. To prevent fissile materials from getting out of control of governments and into the hands of terrorists will require international cooperation on an unprecedented scale to help secure and safeguards such materials wherever they exist. What makes this job so difficult is that less than ten kilograms of plutonium and a few tens of kilograms of highly enriched uranium are required for a bomb, whereas the combined civilian and military inventories exceed a million kilograms.

Finally, nuclear proliferation concerns are exacerbated by the potential expansion of global nuclear power, driven by demands for clean energy and concerns about global climate change. There is considerable disagreement about whether the risks are worth the benefits; nuclear skeptics say no, whereas nuclear advocates believe it the problem manageable, although it requires vigilance. The risk/benefit calculus of nuclear power and nuclear proliferation vary substantially across the globe. The risk must be managed by a combination of technical, political, and institutional measures, whereas the benefits must be measured by how they accrue to individual states as globally.

22

NUCLEAR WEAPON REDUCTIONS MUST BE PART OF STRATEGIC ANALYSIS[1]

HENRY A. KISSINGER AND BRENT SCOWCROFT

A New START treaty reestablishing the process of nuclear arms control has recently taken effect. Combined with reductions in the US defense budget, this will bring the number of nuclear weapons in the United States to the lowest overall level since the 1950s. The Obama administration is said to be considering negotiations for a new round of nuclear reductions to bring about ceilings as low as 300 warheads. Before momentum builds on that basis, we feel obliged to stress our conviction that the goal of future negotiations should be strategic stability and that lower numbers of weapons should be a consequence of strategic analysis, not an abstract preconceived determination.

Regardless of one's vision of the ultimate future of nuclear weapons, the overarching goal of contemporary US nuclear policy must be to ensure that nuclear weapons are never used. Strategic stability is not inherent with low numbers of weapons; indeed, excessively low numbers could lead to a situation in which surprise attacks are conceivable.

---

[1] An earlier version of this essay appeared in *The Washington Post*, Sunday, April 22, 2012.

Perspectives on Complex Global Challenges: Education, Energy, Healthcare, Security and Resilience, First Edition. Edited by Elisabeth Paté-Cornell, William B. Rouse, and Charles M. Vest.
© 2016 John Wiley & Sons, Inc. Published 2016 by John Wiley & Sons, Inc.

We supported ratification of the START treaty. We favor verification of agreed reductions and procedures that enhance predictability and transparency. One of us (Kissinger) has supported working toward the elimination of nuclear weapons, albeit with the proviso that a series of verifiable intermediate steps that maintain stability precede such an end point and that every stage of the process be fully transparent and verifiable.

The precondition of the next phase of US nuclear weapons policy must be to enhance and enshrine the strategic stability that has preserved global peace and prevented the use of nuclear weapons for two generations.

Eight key facts should govern such a policy:

First, strategic stability requires maintaining strategic forces of sufficient size and composition that a first strike cannot reduce retaliation to a level acceptable to the aggressor.

Second, in assessing the level of unacceptable damage, the United States cannot assume that a potential enemy will adhere to values or calculations identical to our own. We need a sufficient number of weapons to pose a threat to what potential aggressors value under every conceivable circumstance. We should avoid strategic analysis by mirror-imaging.

Third, the composition of our strategic forces cannot be defined by numbers alone. It also depends on the type of delivery vehicles and their mix. If the composition of the US deterrent force is modified as a result of reduction, agreement or for other reasons, a sufficient variety must be retained, together with a robust supporting command and control system, so as to guarantee that a preemptive attack cannot succeed.

Fourth, in deciding on force levels and lower numbers, verification is crucial. Particularly important is a determination of what level of uncertainty threatens the calculation of stability. At present, that level is well within the capabilities of the existing verification systems. We must be certain that projected levels maintain – and when possible, reinforce – that confidence.

Fifth, the global nonproliferation regime has been weakened to a point where some of the proliferating countries are reported to have arsenals of more than 100 weapons. And these arsenals are growing. At what lower US levels could these arsenals constitute a strategic threat? What will be their strategic impact if deterrence breaks down in the overall strategic relationship? Does this prospect open up the risk of hostile alliances between countries whose forces individually are not adequate to challenge strategic stability but that combined might overthrow the nuclear equation?

Sixth, this suggests that, below a level yet to be established, nuclear reductions cannot be confined to Russia and the United States. As the countries with the two largest nuclear arsenals, Russia and the United States have a special responsibility. But other countries need to be brought into the discussion when substantial reductions from existing START levels are on the international agenda.

Seventh, strategic stability will be affected by other factors, such as missile defenses and the roles and numbers of tactical nuclear weapons, which are not now subject to agreed limitations. Precision-guided large conventional warheads on long-range delivery vehicles provide another challenge to stability. The interrelationship among these elements must be taken into account in future negotiations.

Eighth, we must see to it that countries that have relied on American nuclear protection maintain their confidence in the US capability for deterrence. If that confidence falters, they may be tempted by accommodation to their adversaries or independent nuclear capabilities.

Nuclear weapons will continue to influence the international landscape as part of strategy and an aspect of negotiation. The lessons learned throughout seven decades need to continue to govern the future.

23

MAINTAINING US LEADERSHIP IN SCIENCE, TECHNOLOGY, AND INNOVATION FOR NATIONAL SECURITY

JACQUES S. GANSLER

For the last six-plus decades, the US national security strategy has been "technological superiority." In the post–World War II years, it was clear that the United States had a significant lead in both commercial and military technologies, as well as in mass production of high-tech goods. However, many recent trends have raised concerns about a potential decline in US technological leadership for future security. Let me briefly list seven key, adverse trends:

1. There has clearly been a globalization of industry and labor, resulting in a global diffusion of technology – leading (in many areas) to a "worldwide convergence of technology."

2. There has been a shift from the historic leadership of US Government investments in R&D, to where US industry (largely commercial) now leads in R&D investments – by a ratio of greater than 2 to 1. And, because of the "isolation" of the US defense industry from the commercial sector (which is due to legislative and regulatory "barriers" to "dual use" operations), combined with the projected, significant

Perspectives on Complex Global Challenges: Education, Energy, Healthcare, Security and Resilience, First Edition. Edited by Elisabeth Paté-Cornell, William B. Rouse, and Charles M. Vest.

declines in DoD budgets, along with the military services' institutional preference for traditional equipment (ships, planes, and tanks), it is likely that the amount of DoD investment in R&D (particularly in longer-term research) is likely to decline even more significantly. At the same time, the overall US government has been largely shifting its science and technology investments from defense to life sciences. Compounding this shift is the fact that the individual firms in the defense industry have been under much greater competitive pressure (due to the declining DoD budgets) so they have shifted more of their dollars and top people to winning near-term proposals versus investing in corporate, long-term R&D.

3. An increasing share of US students (from elementary school, through high school, college, and graduate school) are not choosing science and engineering; and a large share of the graduate students, in US universities, in these fields, are coming from other countries. In addition, 60% of the current US science and engineering workforce will be eligible to retire in the next 5 years. Therefore, there will be a decline in the S&T workforce able to do classified DoD work.

4. Since 2002, the United States has been a net importer of hi-tech goods – with a net trade imbalance of $100 billion last year. Clearly, this is reversing an historic trend – which is consistent with the more recent offshoring of manufacturing jobs, and the increasing emphasis on S&T in many other countries – in full recognition that "the greatest engine of a nation's economic growth is technological change."

5. In numerous areas that are critical to defense, the United States is no longer the world leader, for example, in night-vision devices and in quantum computing. And, when roadside bombs were a major problem in Iraq and Afghanistan, in order to "harden" its vehicles, the DoD looked globally; and chose a South African design for the undercarriage, Israeli armor, German shock absorbers, and some Asian electronics.

6. With limited resources, the military services' "cultural tendency" is to shy away from the "game-changing, disruptive technologies" that are required to truly "stay ahead." For example, two years in a row (when the author was Under Secretary of Defense) the US Air Force did not fund the "Global Hawk" (i.e. no pilot) program, the first US unmanned aircraft. It was clearly counter-cultural; and they had to be directed to fund it; even though it was less expensive and saved lives. Today, unmanned aircraft are acknowledged to be a critical security item.

7. Finally, and related to the prior ("changing culture") trend, is the fact that in 2010, 57% of the DoD's total acquisition dollars went to buying

"services" (vs goods); yet most of the existing government policies, practices, laws, and so on are based on buying goods; and goods are the focus of the government labs; as well as of those in the DoD who decide what R&D to fund. Instead of the sole R&D focus on goods, a significant share of the research and demonstration efforts of the future need to be devoted to "processes" – from providing better and lower-cost services (such as logistics, which is critical for operations and is the most expensive DoD acquisition area), as well as including significantly enhanced manufacturing processes, in order to affordably achieve the required quantity and quality of weapons.

In the last few years, these seven security and economic trends have become widely recognized – for example, in the law, known as the "America COMPETES Act," which was enacted in response to a 2007 National Academies study entitled "Rising Above the Gathering Storm." But there is much resistance, and the recent budget climate has not helped. So, a 2011 National Academies' "revisit" of "the Gathering Storm" report concluded, "the outlook has worsened."

Of course, the optimists note that there are still areas where the United States is believed to be the leader – such as missile defense, robots, silent subs, space, stealth, cyber, large-scale composites, and complex systems integration. But, in many of these areas, the US lead is rapidly disappearing.

So, some near-term actions are required. Specifically, more joint – multinational – R&D, revised US import and export rules on security-related items (in recognition of the adverse trends #1 and #5 mentioned earlier, and take advantage of these "globalization" trends to the benefit of US security); and revised rules on US-based foreign students, employees, and scholars working on unclassified, government-funded R&D (in recognition of adverse trend #3 mentioned earlier, and in order to gain the benefits of these skilled workers for US economic and security needs); and a mandate on an increased share of the DoD budget (say 3%) to be set aside for defense-related research (similar to the mandated set-aside for Small Business Independent Research), with a required focus of these dollars on "disruptive technologies" that are relevant to future security.

In closing, let me emphasize that if the United States is to maintain its national security posture – of "technological superiority" – it must do four things:

1. Think globally (both in alliance security strategy and technology).
2. Be responsive to the changed needs of 21st century security – in terms of what goods and services it buys.

3. Focus on "affordability" – recognition of the nation's economy.
4. Be capable of rapid change (to match the speed of security shifts and technology changes).

These new directions represent a "cultural change," and the literature on successful cultural change is clear; it requires both widespread recognition of the need for change and leadership – with a vision, a strategy, and a set of actions.

The recognition of the need for the change is here. Now, "US leaders" and "world leaders" must take actions to implement this vision.

SECTION V

RESILIENCE

24

INTRODUCTION

Cities have long been a mixture of designed and emergent phenomena. Mumbai, formerly Bombay, became the headquarters of the English East India Company in 1687 and was ruled by the British Raj starting in 1858. Peter the Great founded St Petersburg in Russia in 1703. The British forced the opening of Shanghai as a port for international trade in 1842. All three cities were designed by their rulers to be meccas of modernity. They became gleaming cities amidst seas of poverty. The masses saw these cities as luminous exemplars of all that did not have. Revolution emerged, to the surprise of the industrious designers of these cities (Brook, 2013).

Much more recently in mid-20th century, New York City encountered great difficulties in the 1960s and 1970s. A significant driving force behind these difficulties was a technological innovation – the shipping container (Levinson, 2008). The shipping container slashed the costs of shipping, including virtually eliminating pilferage. New York City had no room to stage and store containers, to New Jersey's benefit. As a result, the City's 1.2 million manufacturing jobs following World War II shrank to less than 100,000 three decades later. Urban designers such as Jane Jacobs (1961) and Robert Moses (Caro, 1974) offered completely different prescriptions, the former focused on urban quality of life and the latter emphasizing the need to better accommodate automobiles.

Moving into the 21st century, the world's population is becoming increasingly urban and many of the largest cities are located near oceans or major

Perspectives on Complex Global Challenges: Education, Energy, Healthcare, Security and Resilience, First Edition. Edited by Elisabeth Paté-Cornell, William B. Rouse, and Charles M. Vest.

rivers. These cities face major environmental threats from hurricanes, typhoons, and storms, the frequencies and consequences of which are aggravated by climate change. As a consequence, there are many major initiatives focused on making cities resilient to such threats (Washburn, 2013).

In parallel, major cities have become more and more politically powerful, particularly as central governments have been stymied by legislative gridlock while mayors have no choice but deliver urban services (Barber, 2013). Cities have clearly become the economic engines that provide countries with resources that are deployed more broadly (Glaeser, 2011). As an example, greater New York City generates 10% of the United States' GDP. In contrast, the less urban states are net consumers of federal resources.

FRAMEWORK FOR URBAN RESILIENCE

Assuring urban resilience involves addressing three problems. First, there is the *technical problem* of getting things to work, keeping them working, and understanding impacts of weather threats, infrastructure outages, and terrorist acts. Developments in urban oceanography provide a good illustration of addressing the technical problem of urban resilience.

Second, there is the *behavioral and social problem* of understanding human, perceptions, expectations, and inclinations in the context of social networks, communications, and warnings. There is a wealth of data and models on human response to natural threats, but as yet our abilities to predict populations' responses have been quite limited (Rouse, 2015)

Third is the *contextual problem* of understanding how norms, values, and beliefs affect people, including the sources of these norms, values, and beliefs. Historical assessments of how the neighborhoods and communities of interest have evolved in terms of risk attitudes, for example, can help to understand how communications should be framed and delivered.

Addressing these three problems requires four levels of analysis:

- *Historical Narrative*: Evolution of the urban ecosystem in terms of economic and social development – What happened, when, and why?
- *Ecosystem Characteristics*: Norms, values, beliefs, and social resilience of urban ecosystem – What matters now and to whom?
- *People and Behaviors*: Evolving perceptions, expectations, commitments, and decisions – What are people thinking and how do they intend to act?

- *Organizations and Processes*: Urban infrastructure networks and flows – water, energy, food, traffic, and so on – How do things work, fail, and interact?

All of this understanding must occur in the context of *this city's* norms, values, and beliefs, as well as its historical narrative. Thus, we must keep in mind that cities are composed of communities and neighborhoods that are not homogenous. This leads us to the levels of analysis of ecosystem characteristics and historical narrative. These levels of analysis provide insights into who should communicate and the nature of messages that will be believed.

To become really good at urban resilience, the following questions need to be addressed:

- How can cities best be understood as a collection of communities and neighborhoods, all served by common urban infrastructures?
- How do policy (e.g., zoning, codes, taxes), development (e.g., real estate, business formation, and relocation), immigration, and so on affect the evolution of communities and neighborhoods within a city?
- When technical problems arise, what message is appropriate and who should deliver it to each significantly different community and neighborhood within the city?
- How can we project and monitor the responses of each community and neighborhood to the information communicated, especially as it is shared across social networks?

Virtual worlds with decision makers in the loop can enable "what if" explorations of these issues and questions. Termed policy flight simulators (Rouse, 2014), such environments can support decision makers and other key stakeholders to understand issues, share perspectives, and negotiate policies and solutions.

POTENTIAL APPROACHES

Batty's recent book, *The New Science of Cities* (2013), epitomizes the application of network theory to social systems. His overall line of reasoning is "Our argument for a new science is based on the notion that to understand place, we must understand flows, and to understand flows we must understand networks. In turn, networks suggest relations between people and places, and thus the central principles of our new science depend on defining relations between the objects that comprise our system of interest."

He continues, "Our second principle reflects the properties of flows and networks. We will see that there is an intrinsic order to the number, size, and shape of the various attributes of networks and thus, in turn, of spaces and places that depend on them, and this enables us to classify and order them in ways that help us interpret their meaning."

These premises lead to Batty's seven laws of scaling that predict the consequences as cities get larger. He reviews a range of models that predict size of cities, income of population, heights of building, locations of retail outlets, and so on over time. The work of Bettencourt also addresses the origins of scaling in cities (Bettencourt, 2013). He defines urban efficiency as the balance between socioeconomic outputs and infrastructural costs. His empirical findings strongly support his theorizing.

Other resources include Pumain (1998) who discusses sources of complexity in urban systems, as well as theories for modeling complexity, drawing mainly on physics and network theory. Van der Veen and Otter (2002) critique models of urban systems and focus on the phenomenon of emerging spatial structure. Zhu (2010) provides a review of computational political science. Gheorghe, Masera, and Katina (2014) outline a broad field of study they term infranomics. In contrast, Byrne (2001) argues for "complexity reasoning" rather than just modeling and simulation and concludes that, "Social structure is more than merely the product of emergent interactions among agents."

OVERVIEW OF CONTRIBUTIONS

Michael Batty, "Urban Resilience: How Cities Need to Adapt to Unanticipated and Sudden Change," argues that cities are the engines of the modern economy in that their wealth appears to increase more than proportionately as their size and influence increases. By and large, most cities are unplanned, growing organically into complex systems that are usually highly adaptable and resilient to various routine crises that threaten to destroy their functioning. However, cities' economies of scale are countered by increased diseconomies due to congestion of various kinds. Various forms of public transport necessarily displace the private automobile. Such systems add to the diversity of ways in which people can interact increasing the redundancy that ensures that routine disruptions can be easily handled. Cities begin to lose their resilience when the connections necessary to their functioning do not keep up with their growth or sometimes their decline.

Globalization of city functions, which has been occurring in the last half century or longer, is destroying this stability. Increasingly cities are specializing in functions that involve networks of trade and exchange, even culture, which are global and thus depend on changes and decisions well beyond the traditional physical city boundaries. The local networks that compose the city and act to boost its resilience to local changes and disruptions are being connected to national and international networks that are much more difficult to control and plan for, and which contain the seeds of economic disruption through contagion that spans the globe. Disruptions can occur anywhere in such systems with their impacts on particular cities being quite unpredictable in the face of our ignorance of what is connected to what.

Richard Reed, "Buying Down Risks and Investing in Resilience," reports that economic losses rise due to the increased frequency and severity of weather-related disaster events affect communities – human, economic, environmental, and social – and often disproportionately affect individuals and families who are most vulnerable. Creating a resilient community is in everyone's best interest and the responsibility of individual citizens, private and nonprofit sectors, and government. He adopts the National Academies' definition of resilience, "the ability to prepare and plan for, absorb, recover from or more successfully adapt to actual or potential adverse events." Often resilience is retrospectively gauged when the adverse forces of natural phenomena or human-caused events collide with individuals, families, businesses, and communities. Many times it involves the cascading impacts of damaged critical infrastructure systems such as electricity, transportation, water, and fuel that exacerbate and contribute to the distress. In many instances, these natural and human hazards are reasonably predictable but in some cases they are not. What is certain is that in varying degrees of severity and frequency, adverse events will impact our homes, livelihoods, and communities. So the question is "how do we as individuals, families, businesses and communities choose to deal with this assured cycle of adversity?"

He argues for an approach that employs the lens of the disaster cycle. The disaster cycle is an intentional sequence of preparedness, response, and recovery actions to create more resilient individuals, families, and communities. Preparing and taking active steps to mitigate risks is a down payment on response. Responding to an adverse event involves the immediate triage of impacts and mitigating those consequences to find stability as quickly as possible. Recovery is the opportunity to make choices that invest in resilience by learning, innovating, and adapting to changing conditions and making ourselves better prepared for the next adverse event.

He concludes that creating resilient communities is a challenging and complex work but the structures and terminology that unite the goodwill and action of disparate organizations and individuals is taking hold. Whether it is an elementary school kid learning what to do when a tornado strikes or the infusion of millions of dollars of federal, state, and philanthropic funds into a community devastated by a disaster, the results are the same – individuals, families, and communities making down payments and investments so that they are better able to absorb and recover from the next event while adapting to a constantly changing environment.

Alexandros Washburn, *"Resilience from the Perspective of a Chief Urban Designer,"* discusses how urban designers make plans to transform cities in anticipation of change. His New York City team had long been working to anticipate the effects of possible hurricanes. They built a digital model of a composite waterfront neighborhood that was realistic but fictitious. Employing a digital model of a hurricane, they analyzed the day-by-day effects on the buildings, infrastructure, and people. They wrote a policy and operations playbook, imagining recovery as a rapid process of transformation, an urban design process.

The reality of Hurricane Sandy overwhelmed their plans. Everything having to do with recovery was much more complicated in real life than in their simulation. Instead of one level of government trying to take the lead, there were three. The city, state, and federal agencies put out competing plans. Instead of one time frame for recovery, there were three: an emergency recovery time frame just to get peopled sheltered before winter, an interim housing and coastal protection time frame that could last years, and a permanent time frame to build a successful shield against climate change that will need to endure well into the century.

Amidst this confusion, there was a need for one level of government to take the lead. He believes that municipal government is the best placed to act and proposes land-use regulation as the best lever. This is one area of law where a city can act unilaterally without waiting for state or federal government. Changes in land use coupled to public space improvements can make a neighborhood resilient. For example, because coastal protections such as berms and sea walls can double as public parks, there is precedent for funding them as public space improvements.

Theo Toonen, "Engineering for Resilience," articulates the "ten commandments" of the Dutch approach to keeping their feet dry. He notes that the key goal of modern Dutch water engineering has always been prevention; not the ability to quickly bounce back after the fact of a flood-related emergency. Consequently, in the Dutch system, there was no real perception and understanding of the scale and enormous consequences of flooding for high-tech economies and highly urbanized societies. However, events in the United States such as Katrina and Sandy have prompted the recognition that resilience has to be part of a broader institutional strategy.

He presents and discusses ten critical institutional design principles of the contemporary Dutch approach to the governance of water management. These ten principles reflect the physical, economic, social, and political complexity of preparation, response, and recovery in the face of enormous challenges. These principles appear to be readily applicable to different political systems, countries, or cultures, as well as different sectors or domains.

REFERENCES

Barber BR. *If Mayors Ruled the World: Dysfunctional Nations, Rising Cities*. New Haven, CT: Yale University Press; 2013.

Batty M. *The New Science of Cities*. Cambridge, MA: MIT Press; 2013.

Bettencourt LMA. The origins of scaling in cities. Science 2013;340:1437–1440.

Brook D. *A History of Future Cities*. New York: Norton; 2013.

Byrne D. What is complexity science? Thinking as a realist about measurement and cities and arguing for natural history. Emergence 2001;3(1):61–76.

Caro RA. *The Power Broker: Robert Moses and the Fall of New York*. New York: Knopf; 1974.

Gheorghe AV, Masera M, Katina PF, editors. *Infranomics: Sustainability, Engineering Design and Governance*. London: Springer; 2014.

Glaeser E. *Triumph of the City: How Our Greatest Invention Makes Us Richer, Smarter, Greener, Healthier, and Happier*. New York: Penguin; 2011.

Jacobs J. *The Death and Life of Great American Cities*. New York: Random House; 1961.

Levinson M. *The Box: How the Shipping Container Made the World Smaller and the World Economy Bigger*. Princeton, NJ: Princeton University Press; 2008.

Pumain D. Urban research and complexity. In: Bertuglia CS, Bianchi G, Mela A, editors. *The City and Its Sciences*(Chap. 10). Heidelberg: Physica Verlag; 1998.

Rouse WB. Human interaction with policy flight simulators. J Appl Ergon 2014;45(1):72–77.

Rouse WB. *Modeling and Visualization of Complex Systems and Enterprises: Explorations of Physical, Human, Economic, and Social Phenomena*. Hoboken, NJ: Wiley; 2015.

Van der Veen A, Otter HS. Scale in space. Integr Assess 2002;3(2–3):160–166.

Washburn A. *The Nature of Urban Design: A New York Perspective on Resilience*. Washington, DC: Island Press; 2013.

Zhu L. *Computational Political Science Literature Survey*. State College, PA: College of Information Sciences and Technology, Pennsylvania State University; 2010.

25

URBAN RESILIENCE: HOW CITIES NEED TO ADAPT TO UNANTICIPATED AND SUDDEN CHANGE

Michael Batty

Cities are quintessentially places where people come together to pool their ideas and resources, to engage in trade, to exchange, and to interact socially and culturally so that their collective prosperity is enhanced. They are the engines of the modern economy in that their wealth appears to increase more than proportionately as their size and influence increases. They are built both from the bottom up and the top down, with individuals and groups putting in place the myriad of connections and networks, which form the infrastructure necessary for their functioning. By and large, most cities are unplanned, growing organically into complex systems that are usually highly adaptable (Batty, 2005). In this sense, they are resilient to various disruptions in that their networks provide a degree of redundancy so that they are able to deal with routine crises that threaten to destroy their functioning.

As cities grow, however, their economies of scale, which are usually called agglomeration, are countered by increased diseconomies due to congestion of various kinds. In very large cities with populations more than about 3 million, the physical modes necessary to move people around efficiently are essentially provided collectively using various forms of public transport and in cities larger than this, the private automobile becomes less and less efficient to use. Indeed, cities that grow rapidly often find it difficult to adapt their

Perspectives on Complex Global Challenges: Education, Energy, Healthcare, Security and Resilience, First Edition. Edited by Elisabeth Paté-Cornell, William B. Rouse, and Charles M. Vest.

physical transportation networks to enable effective functioning, and in the fastest growing cities whose prosperity is also increasing dramatically such as those in East Asia, new public transit systems in the form of high-speed rail and subway systems are fast being put in place to keep such cities working. Such systems combined with other forms of transport add to the diversity of ways in which people can interact increasing the redundancy that ensures that routine disruptions can be easily handled.

Cities begin to lose their resilience when the connections necessary to their functioning do not keep up with their growth or sometimes when their economies decline in such a way that their transport systems are unable to support a declining population, as for example in some rustbelt cities of Europe and the United States. However, this way of looking at urban resilience is strongly conditioned by the notion that the networks that make up cities are well defined and relatively stable. Cities of course reach out into their hinterlands drawing in resources and populations but traditionally acting in synergistic ways to maintain a degree of economic resilience that is stable. Globalization of city functions, which has been occurring in the last half century or longer, is destroying this stability. Increasingly cities are specializing in functions that involve networks of trade and exchange, even culture, which are global and thus depend on changes and decisions well beyond the traditional physical spheres of influence that we associate with any city: beyond physical city boundaries.

In short, the local networks that compose cities and act to boost their resilience to local changes and disruptions are being connected to national and international networks that are much more difficult to control and plan for, and which contain the seeds of economic disruption through contagion that spans the globe. This has not only been made possible by advances in physical transportation technologies but by developments in information technologies that underpin an increasingly large proportion of economic activity in cities and which tend to be largely invisible to our attempts to track what their implications are for a city's local economy. Disruptions can occur anywhere in such systems with their impacts on particular cities being quite unpredictable in the face of our ignorance of what is connected to what. The recent credit crunch and the Great Recession, which began in 2007, have their seeds in such global contagion. Similar contagion can now occur rapidly with respect to the transmission of diseases such as bird flu, even Ebola, due to the movement of populations through air travel, illustrating once again the all-pervasiveness of disruptive forces that spread in systems where everything is connected to everything else, at least indirectly, if not directly.

In the context of how populations are connected to one another in cities, urban resilience takes many forms. First, there is the possibility that transport systems might be disrupted due to technical failures in track and stock as well as events such as terrorism or medical alerts. Usually, in well-adapted cities where the provision of transport infrastructures has great redundancy, travelers can divert and find alternative ways of moving to their destinations but this depends on the severity of the event. For example in early July 2012, the Circle and District lines were closed in London for 4 hours and it was calculated that 1.23 million passengers on the entire subway system were disrupted in terms of their travel time. Much longer-term physical disruptions due to climate change, for example, can pose very severe problems for cities. The IPCC predict that the North Sea will rise by 1 m in the next 100 years and if nothing were done, the Thames Estuary would flood and all the bridges east of Hammersmith would be under water, thus massively disrupting traffic patterns and effectively moving the center of London 3 miles to the west. Of course, this prospect is simply a scenario but it gives some sense of how resilient are existing locations of employment and housing and the transportation connecting them.

Much more complex in terms of urban resilience are the threats posed by an increasingly connected global economy underpinned by extensive information technologies which in themselves provide massive network redundancy. When Lehman Brothers collapsed in New York City in 2008, images of redundant workers in the London Docklands being escorted out of their offices there are enduring. The position of the world's key financial centers in places such as the City of London and Wall Street are ever more vulnerable as finance moves to the net while open labor and capital markets that have been fast developed during the last 20 years are distorting the housing markets of major world cities where indigenous populations are being locked out of affordable living and working conditions. The upside of increased connectivity in cities is redundancy with respect to how urban functions become ever more resilient but the downside is the fact that such increased connectivity allows the diffusion of unanticipated and undesirable forces from physical to financial diseases. We require a much better understanding of such effects and influences with respect to how resilient are cities are in these terms and much of this must come through understanding the complexity of the artifacts of the contemporary urban world that we continue to create (Batty, 2013).

REFERENCES

Batty M. *Cities and Complexity: Understanding Cities with Cellular Automata, Agent- Based Models, and Fractals*. Cambridge, MA: The MIT Press; 2005.

Batty M. *The New Science of Cities*. Cambridge, MA: The MIT Press; 2013.

26

BUYING DOWN RISKS AND INVESTING IN RESILIENCE

RICHARD REED

Creating resilient communities is a challenging and complex work. As economic losses rise due to the increased frequency and severity of weather-related disaster events, there is a toll in our communities – human, economic, environmental, and social – that often disproportionately affects individuals and families who are most vulnerable. At the American Red Cross, it is our mission to *"prevent and alleviate human suffering in the face of emergencies by mobilizing the power of volunteers and the generosity of donors."* Creating a resilient community is in everyone's best interest. There is equal responsibility and work to be accomplished for individual citizens, the private and nonprofit sectors, and government.

While there are a number of definitions for resilience, the National Academies defines it as "the ability to prepare and plan for, absorb, recover from or more successfully adapt to actual or potential adverse events" (NAE, 2012). Often resilience is retrospectively gauged when the adverse forces of natural phenomena or human-caused events collide with individuals, families, businesses, and communities. Many times it involves the cascading impacts of damaged critical infrastructure systems such as electricity, transportation, water, and fuel that exacerbate and contribute to the distress. In many instances, these natural and human hazards are reasonably predictable

Perspectives on Complex Global Challenges: Education, Energy, Healthcare, Security and Resilience,
First Edition. Edited by Elisabeth Paté-Cornell, William B. Rouse, and Charles M. Vest.
© 2016 John Wiley & Sons, Inc. Published 2016 by John Wiley & Sons, Inc.

but in some cases they are not. What is certain is that in varying degrees of severity and frequency, adverse events will impact our homes, livelihoods, and communities. So the question is: how do we as individuals, families, businesses, and communities choose to deal with this assured cycle of adversity?

At the Red Cross, we approach this question through the lens of the disaster cycle. The disaster cycle is an intentional sequence of preparedness, response, and recovery actions to create more resilient individuals, families, and communities. Preparing and taking active steps to mitigate risks is a down payment on response. It is buying down the monetary and emotional expenses that you will pay at some point in the future when confronted with an adverse event. This could be installing smoke detectors and rehearsing evacuation plans for home fire hazards; defining and rehearsing how to move to a safe area or constructing a safe room in your home for tornado and severe weather risks; or completing an all-hazards preparedness plan and kit that provides for basic health, safety, and administrative items and allows you to absorb the financial, emotional, and administrative shocks that accompany any type of disaster. The down payment on response alleviates the monetary and emotional impacts when the event occurs.

Responding to an adverse event involves the immediate triage of impacts and mitigating those consequences to find stability as quickly as possible. The more down payments made while preparing for adverse events, the more quickly individuals, families, businesses, and communities regain the ability to help themselves. Stability allows them to proceed through the myriad activities, including learning and adapting, that return them to the disaster cycle more robust than before.

In some instances, down payments on response are not made or are insufficient to mitigate the scope of shock delivered by the adverse event. In these situations, individuals, families, businesses, or communities cannot stabilize themselves. They need external help to get stabilized and will typically require additional assistance to fully recover from the event. In the disaster cycle, recovery is the opportunity to make choices that invest in resilience by learning, innovating, and adapting to changing conditions and making ourselves better prepared for the next adverse event.

In addition to a simple model that focuses on the cyclical renewal required to become more resilient to disasters, organizations that work with communities must bring clear operating principles, consistent with their mission, to guide the complex and interactive work of making communities more resilient. The principles not only serve as a foundational guide for the organization providing the help but also benefit the recipients by clearly articulating the values and principles behind the services being rendered. At the American Red Cross, our clients and communities can expect the following:

- Services will span the entire disaster cycle (preparedness, response, and recovery) and that those services will be *predictable and repeatable* and applied consistently across the country. The community knows what to expect from us.
- Services and programs are designed first on the *needs and expectations of clients and communities consistent with the mission*, and then on those of key stakeholders.
- We will be a *facilitative leader* across the disaster cycle. The Red Cross will align with local, state, tribal, and federal government entities and work to enable the entire community to participate in all phases of the disaster cycle by being not only a provider of direct services but also a facilitative leader.
- We will use *a single integrated approach* to building personal and community resilience that encompasses services delivered through a comprehensive disaster management process (preparedness, response, and recovery), which integrates and unifies programs and activities *across the entire enterprise.*
- Our organization and culture *continually innovate* in response to client and constituent needs.
- The *speed and accessibility* of our services meets the urgent needs of our clients.

In 2011, Presidential Policy Directive/PPD-8 provided the overarching direction required for an "all-of-Nation" approach to achieve *National Preparedness* (PPD, 2011). While the policy is only directive for federal departments and agencies, it, and the supporting family of frameworks and operational plans, provides the necessary structure and terminology for individuals, private and nonprofit organizations, public utilities, academic institutions, and government entities at all levels to align and galvanize their activities toward unified national preparedness outcomes. Our federal form of government combined with individual citizen activism, the interests of public and private utilities, and the widely varying role of private and nonprofit organizations make coordinating preparedness efforts challenging. But our collective awareness and understanding of what is required to build resilient individuals and communities is improving. Retaining a focus on the cyclical and certain nature of adverse events and working with others who share common goals to make individuals and communities more robust for the next event is critical for success.

The Red Cross was proud to work with hundreds of organizations and individuals to rally around the citizens of Moore, Oklahoma after the devastating EF-5 tornado in May 2013. That was one of the first events where we put into practice the disaster cycle concepts and principles outlined earlier. Two years later, we continue to learn and evolve our cyclical approach for helping individuals and communities make down payments on response and investments in preparing for the next likely event. One example is our partnership with the Walt Disney Company, where we teach preparedness to children through the simple and comfort-giving medium of a pillowcase. After a tornado struck Moore again this year, the following message was posted to the local Red Cross Facebook page: "While we were in the shelter today my daughter was quoting things she learned from you guys yesterday (in a pillowcase preparedness presentation at the elementary school). She was calm, a big difference from last year. What you do makes a difference. Thank you." Regardless of how small or inconsequential the down payment on response may seem, it can mitigate enormous physical and emotional expenses at a later date.

At the community end of the spectrum, the Department of Housing and Urban Development, in collaboration with the Rockefeller Foundation, has created an innovative National Disaster Resilience Competition to award $1 billion to communities devastated by disasters in recent years. "The competition will help communities recover from prior disasters and improve their ability to withstand and recover more quickly from future disasters, hazards and shocks" (HUD, 2014).

Creating resilient communities is a challenging and complex work but the structures and terminology that unite the goodwill and action of disparate organizations and individuals is taking hold. Whether it is an elementary school kid learning what to do when a tornado strikes or the infusion of millions of dollars of federal, state, and philanthropic funds into a community devastated by a disaster, the results are the same – individuals, families, and communities making down payments and investments so that they are better able to absorb and recover from the next event while adapting to a constantly changing environment.

REFERENCES

HUD (2014). *"National Disaster Resilience Competition Fact Sheet,"* Washington, DC: Department of Housing and Urban Development. Available at www.portal .hud.gov/hudportal/documents/huddoc?id=NDRCFactSheetFINAL.pdf.

NAE. *Disaster Resilience A National Imperative.* Washington, DC: The National Academies Press; 2012.

PPD. 2011. Presidential Policy Directive – 8, National Preparedness. Available at www.dhs.gov/presidential-policy-diretive-8-national-preparedness.

27

RESILIENCE FROM THE PERSPECTIVE OF A CHIEF URBAN DESIGNER

Alexandros Washburn

Urban designers make plans to transform cities in anticipation of change. As Chief Urban Designer for New York City under Mayor Michael Bloomberg, my team had been working with the Office of Emergency Management since 2007 to anticipate the effects of a possible hurricane strike. Yet nothing in our planning could prepare us for the full reality of Hurricane Sandy in 2012.

We had tried not to be alarmist in the early days. We had built a digital model of a composite waterfront neighborhood that was realistic but fictitious, naming it "Prospect Shore." We hit it with a digital model of a hurricane, and analyzed the day-by-day effects on the buildings, infrastructure, and people of Prospect Shore, represented by several avatars of the imagined community. We wrote a policy and operations playbook, imagining recovery as a rapid process of transformation, an urban design process.

In our computer scenario, we imagined that the old brick buildings in the neighborhood that survived the storm were hardened with reinforcing steel and integrated into new hurricane-resistant building structures. The new buildings would have ground floors reserved for retail and parking. The geometry of their envelopes would be optimized to withstand wind forces. The new streets and buildings in Prospect Shore would drain into a restored

Perspectives on Complex Global Challenges: Education, Energy, Healthcare, Security and Resilience, First Edition. Edited by Elisabeth Paté-Cornell, William B. Rouse, and Charles M. Vest.
© 2016 John Wiley & Sons, Inc. Published 2016 by John Wiley & Sons, Inc.

waterway where native plants would bioremediate any toxins before they reached the harbor.

Industry would be rebuilt at the waterfront but would be integrated with a public esplanade protected from storm surge by a planted earthen levee that doubled as a park. Wetlands would take root offshore and a barrier island in the harbor would become an oyster habitat. We imagined in our computer scenario that we succeeded. Residents and new neighbors returned to find Prospect Shore physically better prepared for storms, socially better integrated into the surrounding city, and therefore more sustainable and resilient than before.

The reality of Hurricane Sandy overwhelmed our imaginings, both personally and professionally. I live in the waterfront community of Red Hook, Brooklyn, and I defied the mayor's evacuation order for Hurricane Sandy because I wanted to see first hand whether the storm surge would turn Red Hook into Prospect Shore. The storm surge that inundated us on October 29, 2012 met our expectations for destruction, but the path to recovery was not so clear.

Everything having to do with recovery is at least three times more complicated in real life than in our simulation. Instead of one level of government trying to take the lead, there are three. The city, state, and federal agencies put out competing plans. Instead of one time frame for recovery, there are three: an emergency recovery time frame just to get people sheltered before winter, an interim housing and coastal protection time frame that could last years, and a permanent time frame to build a successful shield against climate change that will need to endure well into the century.

The third complication is how to decide which scale to adapt. Alternatives include the individual building scale, where each homeowner foots the bill; the regional scale where the federal government pays to protect an entire coastline; or somewhere in between where some combination could protect a neighborhood. The risks and costs of protecting what we have are enormous and there is no consensus.

The federal government is trying to put out new flood maps. While the mapped flood elevations change, the city is trying to process building permits based on shifting data. Meanwhile, private insurance companies are redlining any neighborhood within 3,000 ft of water.

Amidst this confusion, I see the need for one level of government to take the lead, and I believe that municipal government is best placed to act with respect to New York City's resilience. The silver bullet is land-use regulation, the one area of law where a city can act unilaterally without waiting for state or federal government. If changes in land use coupled to public space improvements could make a neighborhood resilient, then there is hope for recovery.

For example, because coastal protections such as berms and sea walls can double as public parks, there is precedent for funding them as public space improvements.

And in a city where there is demand for more density, changes in land use that provide more density generate enormous resources. Such changes can be coupled to public space improvements. The classic New York City example is the High Line project that transformed The West Chelsea neighborhood of Manhattan from a derelict meat packing district to one of the most sought-after mixed-use neighborhoods in the world. The economics were simple: $100 million of public money to build the High Line park coupled with a rezoning for more density of the surrounding neighborhood led to $3 billion of private investment. The same sort of leverage could be used to rebuild and make resilient neighborhoods such as Red Hook.

I would like to say that the only challenge on the path to resilience is urban design's classic dilemma: how to achieve growth while preserving character. But perhaps I misspoke when I said that land-use regulation is unilateral. There is one veto of land-use regulation the state and federal governments retain: environmental review. Charged with protecting nature from us, wouldn't it be ironic if environmental law stopped cities from protecting us from nature?

28

ENGINEERING FOR RESILIENCE

Ten Commandments of the Dutch Approach

THEO TOONEN

Resilience is hot. The Netherlands claims the status of "safest delta in the world." The OECD speaks of "an excellent track record on water management" (OECD, 2014). Yet, resilience has not been one of the central design principles of Dutch water management. The key goal of modern Dutch water engineering has always been prevention not the ability to quickly bounce back after the fact of a flood-related emergency.

For a long time, neither lay people nor professional engineers wanted to have to think in terms of "resilience." Rather, the event of the emergency – floods, breaking dikes, "wet feet" – had to be prevented. Government was considered responsible. Discussing the fact that things could go wrong, that people should prepare themselves in case of emergency, only raised the question of whether the government had not properly done its job. Ministers appearing on television, telling viewers they would need an emergency kit to survive the first few days after the flood or, even worse, that perhaps it would be wise to store a rubber rescue boat at the attic – "just in case" – were ridiculed. Questions in parliament: "Is this to say that the Netherlands is not save anymore against flooding?" Was the country in danger? Did the government fail one of its most crucial jobs – safety? Resilience sounds good. But it is a multilayered concept. It easily becomes contested.

Perspectives on Complex Global Challenges: Education, Energy, Healthcare, Security and Resilience,
First Edition. Edited by Elisabeth Paté-Cornell, William B. Rouse, and Charles M. Vest.
© 2016 John Wiley & Sons, Inc. Published 2016 by John Wiley & Sons, Inc.

The average Dutchman did not and does not pay much attention to water safety. It is considered a given. Not God given, but coming close. Water safety is still generally considered to be the responsibility of water experts – engineers – and government to take care. If something goes wrong, which occasionally occurs, blaming the government is the norm. Even the famed Dutch water management system is not without its flaws. Small-scaled incidents result in much public attention. This actually seems part of the systems strategy to make itself resilient, that is, deal with the consequences of emergencies and learn from it to improve overall system performance.

There is an inherent paradox in resilience thinking. Some water experts – not only jokingly – maintain the need for a water- or flood-related incident every now and then: "A small flooding every odd day to keep the real disasters away." In a way and as a system, the Netherlands in recent times has learned more from international than local knowledge and experience. It was Hurricane Katrina (2005) – and later super storm Sandy (2012) – that unequivocally drove home the message to a self-assured and self-proclaimed body of experts and policy makers. In the Dutch system, there was no real perception and understanding of the scale and enormous – first-, second-, third-order, and cascading – consequences of a flooding for our high-tech economies and highly urbanized societies.

From a resilience engineering point of view, the country had been "too safe." People were surprised about the amount of time and effort it required the New Orleans region to recover. "One would expect this in Bangladesh. Not in a highly developed country like the US." Also the gravity of the social consequences became tangible. Suddenly, at least some people realized that also in the Netherlands many of the poorer neighborhoods are located like bathtubs below the water levels of adjacent rivers and canals running through their cities. When talking about resilience, water presents itself not only as a technical, but also as an organizational, administrative, social, and political issue.

Resilience, ultimately, is a systemic characteristic. It is not an individual tool, technique, or decision. Abilities to engineer for resilience should be studied at the level of engineering systems, next to engineering technology, protocols, and artifacts. At the system level, resilience quickly turns into the questions of learning, adaptation, and survival. In engineering for resilience, the focus of attention has to be the link between functional, operational features on the one hand and systemic characteristics on the other (Toonen, 2010). This link is the world of interdependencies, governance, and the interplay of technology, institutions, and human behavior. Resilience engineering has to be part of a broader institutional strategy. At the operational level, resilience entails the capacity to bounce back. At systems level, resilience is about adaptation and survival. It entails a mobilization of a bias toward the recognition of unanticipated – leave alone undesired – consequences and the ability to take them into account before they actually have happened. In short, to "be prepared for change" (Dietz, Ostrom, & Stern, 2003).

The contemporary Dutch water management system has a long historical development, dating back to well before the 12th century (Toonen, Dijkstra, & van der Meer, 2006). Unlike the popular belief in the infallibility of the Dutch water management system, its long-term history is a running account of many larger and smaller catastrophic events that have more or less successfully been overcome (Dolfing, 2000). The popular American saying, for example, that "God created the world, only the Dutch created their own country" over-looks the fact that – a long time ago – to a large extent the Dutch created God's problem first. In large-scale digging for peat in order to heat homes and premises, the Hollanders had created themselves a flood-prone habitat. But indeed, seen over a longer stretch of time and despite larger and smaller incidents, the Dutch did not witness anything one would these days call a "system failure." Incidents and disasters often have provided the interlude to the improvement of the system. *Rebuilt by design*. Seen this way, the Dutch system of water governance perhaps has a better proven-record of resilience, than of prevention.

The prominent role of government in Dutch water management could merely be considered as yet another example of "European socialism." This could be unacceptable to a different ideological culture or climate. Therefore, it might not be of much use as an enlightening "best practice" to other countries. I would maintain, however, that the Dutch water management system rather than a governmental or state run system, may be better perceived and understood as a case of *high trust organization*. These do not emerge out of the blue. Similar to water management, they are manmade, usually over a long time. They also need maintenance.

One of the foremost weaknesses – next to certain self-congratulatory and inward-looking tendencies – of high trust systems is, precisely, that the system as a system is easily taken for granted. The average civil engineers' belief and conviction is that centralization in the hands of a few experts is the demonstrated best way to govern the system – any system. In addition, a trusted system causes few people to care for the common good and take a deeper look into the black box of Dutch water management or, more precisely, the nature of the governance of Dutch water management as key to resilience.

More focus is needed. Ten critical institutional design principles of the Dutch approach to the governance of water management may well be applicable to different political systems, countries, or cultures, but also to different sectors or domains (Dietz, Ostrom, & Stern, 2003):

1. a deliberate *systems approach* in assessing and setting adequate boundaries – at functional, spatial, and temporal scales – which requires attention and tolerance *for multiscaled governance*;
2. a solid and independent scientific *knowledge management* coupled to a *public problem analysis*;
3. a polycentric, *multiactor governance* structure based upon participation of markets, stakeholders, and relevant communities;
4. a *co-governance approach* based upon broad opportunities for interest representation and low threshold *mediation* and clear attention for *conflict resolution*;
5. an *international* and *best available technology* (BAT)-based strategy;
6. not centralization, but a *clear, transparent and directive national policy*, which serves as a foundation and constraint for regional ecological, social-economic, and cultural-historic territorial agendas;
7. a differentiated but transparent regional administrative structure, under unequivocal, externally legitimated and preferably independent executive leadership;
8. *institutionalized engagement, stewardship, and participation* where the important and, for strategic problems, relevant regional governmental and nongovernmental "caretakers" and executive organizations are involved;
9. *shared services* for maintenance and strategic asset management for the efficiency of the executant organization as a whole;
10. *Checks and balances*, independent evaluation, inspection, and supervision with an institutionalized capacity for learning, and innovation *for adaptive governance and policy-making*.

SYSTEM OF SYSTEMS

Over the past 20–25 years the structure and process of Dutch water governance underwent considerable changes. Step by step, the system has been reformed and redesigned. It took the threat of near river floods, but the reform was well before the public debate on global warming and climate change actually took off in the Netherlands. The movement was from flood risk analysis toward *water safety systems*. River basement and catchment areas became primary units for demarcating adequate water system boundaries.

It constituted the basis, for example, of a large-scale reform of the Dutch water boards. In addition, no longer individual and separately surveyed and administered stretches of dykes, but "dyke rings" were introduced as the relevant units for adequately setting boundaries of and standards for safety systems. Individual Dutch cities are relatively small by international standards. But *Dijkring 14* suddenly revealed itself as a true Metropolis. A large part of the Western Randstad region – Amsterdam, The Hague, Leiden, Rotterdam – is located within or affected by one and the same "ring" – system – of dykes. This "water safety region" suddenly hosted 4–5 million people. The potential danger did not only come from the sea, but also from the hinterland – the rivers. Would people in case of an emergency get stuck between these opposing forces? Compartmentalization strategies could benefit some cities. But, it would increase the risk inflicted upon others. One had to face the awkward and hitherto unarticulated question how to evacuate a "safety system" this size.

PUBLIC EXPERTISE

Water management in the Netherlands has always been considered a clear case for water experts. In the 1990s, the economics of safety were embraced. The overarching goal for Dutch flood policy of the 21st century became: "Maintaining flood risks at an acceptable level against acceptable costs." Risk became defined as the chance of occurrence times the costs of the consequences of the occurrence. Various measures may affect the risk estimate. Modern risk-management in the Netherlands implies a trade-off between the – estimated and perceived – cost and benefits of the management of the probability of flooding and of the management of the consequences of flooding.

Trade-offs imply value judgments: economic, political, moral. The societal sensitivity of the subject matter calls for solid and independent scientific knowledge management. But this should be combined with a public debate and problem analysis. Not only for reasons of legitimization, but also for effectiveness. The challenge is to translate large-scale, complex, and abstract issues of "global warming" and "climate change" in terms of consequences for people and affected populations as well as the courses of action open to them (Reed et al., 2010). For adaptive – strategically incremental and resilient – systems it is important to break down, disaggregate, and "localize" the overall debate on safety into comprehensive and manageable components. Engineering for resilience focuses on impact and consequences. The methodology may well provide a contextual strategy to incorporate unanticipated outcomes and enhance resilience and adaptive capacity. All this assumes that the water resilience expert meets the "citizen scientist," and vice versa.

CROWDED HOUSE

If anything, the Dutch case illustrates that engineering for resilience requires tolerance for and acceptation of duplication and overlap (Landau, 1969). Dealing with a system of spatial and functional interdependency requires the recognition that we are dealing with the "Law of requisite variety" – local, regional, national, transnational, supranational – of functional and territorial scales. The system has to be "vertically integrated." Management of the modern urbanizing Delta, which the Netherlands is, requires no longer only the involvement of the traditional "water authorities" – in the Netherlands the water boards, Provinces, and *Rijkswaterstaat*. The development over the last 20 years has brought communities and municipalities – drinking water, drainage, and sewage systems – at the local level more central to the field. It introduced various new players at the European regional level, including river basement authorities of various sorts and status – representing the Rhine, Maas, and Scheldt rivers. This also includes the European Commission, giving body to the environmental policies jointly agreed upon to by the EU-member states, through the implementation and enforcement of various "Water Directives."

A resilient system also has to become "horizontally integrated." Resilience presupposes interdisciplinarity. Water management has to be taken out of its policy silo. In the Netherlands, efforts to create *Room for the Rivers* implied the incorporation of Environmental Affairs and Integrated Spatial Planning into the formerly more insulated field of water management. The ability to simulate, comprehend, visualize, and deal with variety – rather than being baffled by the institutional complexities of multiscaled governance – is an essential ingredient for resilience engineering.

CO-GOVERNANCE

In dealing with the emergent institutional complexity, the number of players and veto-points has been reduced. At the turn of the millennium, in little over a decade, the number of water boards in the Netherlands has been spectacularly reduced from far above thousand to a handful of about twenty. Many of these reforms were initiated bottom up. Further reduction of the number to maybe even four or five is not to be excluded. New players, however, entered the field. Dutch water management became a target of innovation and international business strategies. It brought new entrepreneurial spirit to the system. Attention for crossovers and reliance inspired new and innovative concepts: multifunctional dykes, building with nature, urban "water squares." It gave business an even more prominent role in Dutch the water management system than the international engineering companies and globally operating dredging firms already occupied.

The need for "simplification and streamlining" of the overall system has to be found in reducing transaction costs and improving the situational capacity for collaborative management and co-governance across markets, stakeholders, and relevant communities. Interest representation, low threshold mediation, and clear attention for public debate and conflict resolution seem rather dull administrative, legal, and even "bureaucratic" issues. They have to be part of any institutional resilience strategy for it to be effective.

CLEAR DIRECTION

The Great Flood of 1953 resulted in 1836 victims and great economic and regional damage that created a traumatic experience in collective Dutch memory. "This may never happen again." Crisis induced centralization. The aftermath of the disaster in the southwestern part of the country triggered the launch of the famous Dutch Delta Works, the longtime pride and glory of Dutch – not in the least Delft – civil engineering. The Delta Works were initiated by the Dutch government and effectively lead by *Rijkswaterstaat* as its "hands on" executant agency. This has strengthened public impression and engineers' belief systems that water safety is primarily an affair of professionals and served by a strong national government and centralized policy.

The institutional development since the 1990s took a different direction. With the broadening of the scope of Dutch water policy, the role of central government has changed. Two decades of gradual and negotiated reform were consolidated into a National water Act in 2009. This stood for a much broader development. One of them was to improve the strategic capabilities of national authorities. From the executant protector against flooding national government became responsible as the leader of the system for safety. Engineering for resilience needs clear, transparent, and directive national policy. This serves as the foundation and constraint for regional ecological, social-economic, and cultural-historic territorial agendas.

With the water policy for the 21st century – WB21 – program of the late 1990s, simple but well thought out policy design principles were being introduced for guidance. "(1) Delay, (2) Store, and (3) Discharge." It became one of the famous triplets on how to deal with floods caused by rising rivers due to upstream rainfall. Introduced by Dutch central government policy in the 1990s, these principles now serve to design and redesign local government water-safety policies. Not only in the Netherlands, but across the world; from the Vietnamese River Delta's to the City of Hoboken (NJ) across from the former New Amsterdam, on the shores of the Hudson River in the wake of the Sandy flashfloods. Without centralization the impact of national government leadership on local government can have a far reach.

EXECUTIVE LEADERSHIP

Rather than a uniform, centralized executive bureaucracy, modern territorial water management requires a differentiated but transparent regional administrative structure, under unequivocal, externally legitimated, and preferably independent executive leadership. Engineering for resilience, with its inherent trade-offs, requires that "we, the people" become involved. Experience with the Dutch system – outside the urbanized parts of the country – shows that "lay" people are very well aware of the immediate safety circumstances in their local contexts. Yet, functional democracy needs to be juxtaposed with professionalism and vice versa.

In the Netherlands, water boards are constitutionally independent executive organizations. They are the exception to the rule of general interest representation by Dutch governments. They are legally entrusted with the functional task of water management within a designated territory and as such may operate legally from a position of relatively independence vis-à-vis the general interest governments: national, provincial, and municipal.

Individual water boards operate within an enumerated mandate based upon generic law and regulated by bylaws and separately agreed upon policy mandates of provincial and national government. Although, unfortunately, party politics rather than functional politics managed to infiltrate the water board elections, the institutional position still allows for a relatively independent executive and professional operation in the best interest of water safety for their region. As democratically embedded institution, they have the external legitimization to serve the common interest in one of the most fundamental human rights in the Netherlands: water safety. It locates clear institutional responsibility when something goes wrong. The water boards – before the fact – are very aware of their responsibilities and who will eventually get the political and social blame, might something go wrong. Governments, industries, or other "stakeholders" with different priorities or a neglect of the common good cannot easily bypass or overturn them – which does not prevent them from constantly trying.

INTERNATIONAL BAT STRATEGY

Science-based engineering and engineering-based science are key. Professional knowledge and solid science are other key ingredients of engineering for resilience. It has given water management its technocratic – not to mention "dull" – public image. Experts and academic researchers increasingly operate in a global context. Engineering and engineering science has turned into a global industry by a global community of colleagues and competitors. Modern engineering is not a local affair anymore – not even in the Netherlands and whatever the preferences that national corps of engineers in the world might signal to their national governments. Dutch policy is proud to export the success and its – international – experience in water management. It has clearly benefited from importing international experience as well. The Dutch system produces excellent engineers, many of whom acknowledge that as far as technical expertise is concerned international colleagues these days often outperform them.

Engineering for resilience requires an *international* and BAT-based *strategy*. It is important to consider, however, that effective water management implies detailed practical knowledge, knowledge of the job, and knowledge of the circumstances. This knowledge is actually becoming an indispensable part of BAT. Water technology is technology with the human touch. This makes "water knowledge" almost a cultural, rather than a technocratic commodity. Water systems often evoke feelings of regional identity or national pride, sometimes combined with international envy, hampering much-needed cross-national communication. Local knowledge and local presence cannot be missed and are of strategic importance to the system's success. People are part of water technology. Knowledge for engineering for resilience has to combine generic with comparative understanding, physical science with anthropology, and engineering science with humanities.

IMPLEMENTATION DEMOCRACY

Professionally serving the public water interest within its designated territory, water boards are not able to pass the buck and forego institutional responsibility. The relatively independent position and "problem ownership" of water boards may be understood as key factor to the resilience of the Dutch water management system as a whole. An important misunderstanding that needs to be foreclosed for resilience engineering is that the Dutch do pay so much attention to water management because it is of obvious importance to the very existence of their country. The logic of collective inaction is, as we all know, the other way around. If something is important or belongs to everybody, it is important or belongs to nobody, giving rise to all sorts of Dramas of the Commons (Stern et al., 2002). In the Netherlands, water is a collective good. Institutionally, water safety is not easily neglected in the Netherlands. The collective good has historically deliberately been turned into a public good, well to be distinguished from a "governmental" good.

Water management is closely intertwined with the day-to-day households of farmers, firms, and citizens. People and organized interests, who had to carry out the jointly agreed upon policies, in return were represented in the general water governance board within their region. For a long time, legally designated functional interests had a say proportional to their vested interest in the day-to-day water management within the system. The composition of interests and their relative strength in the general board could thus be varied from region to region. This constitutional option to use the composition of governance as a way to put – for example, rural or urban – social reality in coherence with physical reality in different regions and territories became neglected by the provinces. It also became difficult to uphold due to the internal complexities of scale enlargement. Dutch water boards, however, from their origin should be perceived as a form of implementation democracy: those who carry out policies are represented in the governance of those policies. This creates a strong sense of ownership, commitment, and stewardship in the interest of water safety within the given territory.

Dutch democracy is said to have been built upon water. The involvement of different social interests in the governance of water boards actually has a different, a more consociational background. *Waterschap* is the word for "water board" in Dutch. "-*Schap*" historically means something like "joint arrangement." As an executive organization, the history of water boards is that different social interests – farmers, firms, inhabitants, preservationists, citizens – work together in safeguarding their territory. They also have to carry out, within their own day-to-day "household decisions," jointly agreed upon policies that should keep their collective territory safe from flooding. This form of "implementation democracy" has less to do with principles of public representation or popular participation, than with the institutionalization of pragmatism, self-interest and self-governance in compound systems. A critical factor for the sustained resilience of the Dutch system will be the reinvention of this principle, particularly within the urbanized, metropolitan areas of the Western – seaside – part of the "low country."

SHARED SERVICE

Next to its relative independence and stewardship in serving the general interest of water safety for its region, Dutch water boards have historically secured another crucial value for the ability toward resilience engineering: the attention for the maintenance of local and regional water systems. The maintenance and permanent updating of water infrastructures – the daily care for dykes, waterways, and systemically related critical infrastructure from the viewpoint of the interdependent safety-system – or the lack thereof – is a critical factor for the resilience of the system. Engineering for resilience, in essence, is part of strategic asset management. Usually, things go wrong when regular day-to-day maintenance has been neglected for a longer time or could not be afforded. They turn into a disaster if linkages to related social, political, and economic domains have not been kept up to date.

Maintenance of critical infrastructures is costly, politically not spectacular, and socially rewarding, but not very visible. Constant saving for investment has to be secured. Engineering for resilience requires attention for the "back-office" organizational and financial infrastructure in order to have the means available before one has to declare a territory a "disaster" or "emergency area." It is an underestimated domain in the analysis of Dutch water management. Yet, there is hardly a better *insurance* than to keep savings and investment capacity at par with the required costs of maintenance and modernization of water and water related infrastructure.

CHECKS AND BALANCES

Finally, for a system to be resilient, it is necessary to be able to face the consequences of decisions or otherwise be confronted with them well before they actually occur. It is always sensitive to openly discuss the weaknesses of the system. A resilient system assumes room for "whistle blowers." Calls for change against vested interests have to be cherished. Proposals for "innovation" as the way out should be carefully scrutinized in terms of their actual contribution to the safety, resilience, and adaptive capacity of the system. One should not have to wait for the next flood in order to spur the attention for safety and security. In short, however we hate it, surveillance, independent supervision, and professional bureaucracy are essential to the systemic ability to engineer for resilience. Even the Netherlands requires international organizations to remind us of the simple fact that external accountability is the key to resilience.[1]

REFERENCES

Dietz T, Ostrom E, Stern PC. The struggle to govern the commons. Science 2003;302:1907–1912.

Ostrom E. A general framework for analyzing sustainability of social-ecological systems. Science 2009;325:419–422.

Dolfing B. Waterbeheer Geregeld; een historisch bestuurskundige analyse van de institutionele ontwikkeling van de hoogheemraadschappen van Delfland en Rijnland 1600-1800 [PhD dissertation]. Leiden University; 2000.

Landau M. Redundancy rationality and the problem of duplication and overlap. Public Adm Rev 1969;29(4):346–358.

OECD (2014), *Water governance in the Netherlands; fit for the future? OECD Studies on Water*, OECD Publishing 10.1787/9789264102637-en.

Reed WV et al. Earth system science for global sustainability: grand challenges. Science 2010;330:916–917.

Stern PC, Dietz T, Dolsak N, Ostrom E, Stovich S. Knowledge and questions after 15 years of research, National Resource Council. In: Ostrom E, Dietz T, Dolsak N, Stern PC, Stovich S, Weber EU, editors. *The Drama of the Commons. Committee on the Human Dimensions of Global Change*. Washington, DC: National Academy Press; 2002. p. 456.

Toonen T. Resilience in public administration: the work of Elinor and Vincent Ostrom from a PA perspective. Public Adm Rev 2010;70(2):193–202.

Toonen T.A.J., G.S.A. Dijkstra, F. van der Meer, Modernisation and reform of Dutch Water Boards: resilience or change? J Inst Econ, 2, 2, 2006, Cambridge University Press: 181–201.

[1]OECD, op cit, 2014: 243 pp.

CONCLUSIONS

The political polarization that has increasingly affected the United States in the last decades has seriously eroded its ability to handle the trade-offs that it needs to face constructively to maintain the prosperity and the competitiveness of the country. In the end, its success in all the areas discussed in this book will depend on the leadership and the strategic vision of those who are elected and appointed to key government positions. At the same time and on the same level, it will depend on the vision of its industrial leaders who shape the economy, with a sense of innovation and creativity. Public–private partnership has never been more essential to ensure that the economic growth benefits the population at large, and that government and industry work together to find the right balance between regulation and productivity, between high employment levels and workers well-being, and between environmental concerns and the needs of the country.

Many of the US internal problems require a world vision. It starts with cultivating our relations with our partners and our allies, in commerce, diplomacy, cultural matters and defense, supporting our principles while understanding other perspectives. It also means learning from the successes and the challenges of others, respecting what is respectable and fighting what is worth fighting for.

Perspectives on Complex Global Challenges: Education, Energy, Healthcare, Security and Resilience, First Edition. Edited by Elisabeth Paté-Cornell, William B. Rouse, and Charles M. Vest.
© 2016 John Wiley & Sons, Inc. Published 2016 by John Wiley & Sons, Inc.

It will take a new generation of people of stature, men and women, with a balanced and systemic vision and a willingness to confront, in a consistent and global way, difficult tradeoffs in turbulent times. The environment in which our grandchildren will be at the helm is difficult to predict. One can imagine a world in which borders are less and less relevant, or at the other end of a spectrum of scenarios, one in which countries retreat behind their walls as seems to be the trend in some quarters of Europe. The challenge is thus to prepare our children – and the heritage that will be theirs – at a fundamental level, with bases that will allow them to address effectively the problems described in this book (and many more): improved healthcare (and cost control) at a time when life spans expand; education that provides wide access to knowledge, allows the best teachers to thrive and gives the students the freedom to think deeply and differently; an effective defense system that discourages aggressions under various motives; the energy that they need without destroying the environment; and livable and pleasant cities. These problems of course are deeply interconnected and have to be addressed as a system, guided by values and technologies.

We hope that the variety of thoughts presented here will help design balanced policies, in the hands of generous and realistic people.

INDEX

Perspectives on Complex Global Challenges: Education, Energy, Healthcare, Security and Resilience,
First Edition. Edited by Elisabeth Paté-Cornell, William B. Rouse, and Charles M. Vest.
© 2016 John Wiley & Sons, Inc. Published 2016 by John Wiley & Sons, Inc.